PROTO

Also by Laura Spinney

The Doctor
The Quick
Rue Centrale: Portrait of a European City
*Pale Rider: The Spanish Flu of 1918 and
How It Changed the World*

PROTO

How One Ancient Language Went Global

Laura Spinney

WILLIAM
COLLINS

William Collins
An imprint of HarperCollins*Publishers*
1 London Bridge Street
London SE1 9GF

WilliamCollinsBooks.com

HarperCollins*Publishers*
Macken House
39/40 Mayor Street Upper
Dublin 1
D01 C9W8, Ireland

First published in Great Britain in 2025 by William Collins

2

Copyright © Laura Spinney 2025

Laura Spinney asserts the moral right to be identified as the author of this work in accordance with the Copyright, Designs and Patents Act 1988

A catalogue record for this book is available from the British Library

ISBN 978-0-00-862652-5 (hardback)
ISBN 978-0-00-862653-2 (trade paperback)

All rights reserved. No part of this publication may be reproduced, stored in a retrieval system, or transmitted, in any form or by any means, electronic, mechanical, photocopying, recording or otherwise, without the prior permission of the publishers.

Without limiting the author's and publisher's exclusive rights, any unauthorised use of this publication to train generative artificial intelligence (AI) technologies is expressly prohibited. HarperCollins also exercise their rights under Article 4(3) of the Digital Single Market Directive 2019/790 and expressly reserve this publication from the text and data mining exception.

This book is sold subject to the condition that it shall not, by way of trade or otherwise, be lent, re-sold, hired out or otherwise circulated without the publisher's prior consent in any form of binding or cover other than that in which it is published and without a similar condition including this condition being imposed on the subsequent purchaser.

Set in Adobe Garamond Pro by Palimpsest Book Production Limited, Falkirk, Stirlingshire

Printed and bound in the UK using 100% renewable electricity at CPI Group (UK) Ltd

This book contains FSC™ certified paper and other controlled sources to ensure responsible forest management.

For more information visit: www.harpercollins.co.uk/green

For Ryszard

Men go abroad to wonder at the heights of mountains, at the huge waves of the sea, at the long courses of the rivers, at the vast compass of the ocean, at the circular motions of the stars, and they pass by themselves without wondering.

The Confessions of Saint Augustine, c. 400 CE

CONTENTS

PROLOGUE   1

INTRODUCTION   7
Ariomania

1. Genesis   35
Lingua obscura

2. Sacred Spring   63
Proto-Indo-European

3. First Among Equals   93
Anatolian

4. Over the Range   121
Tocharian

5. Lark Rising   145
Celtic, Germanic, Italic

6. The Wandering Horse   183
Indo-Iranian

7. Northern Idyll   213
Baltic and Slavic

8. They Came from Steep Wilusa   237
Albanian, Armenian, Greek

CONCLUSION   259
Shibboleth

FIGURES   277
ACKNOWLEDGEMENTS   279
BIBLIOGRAPHY   283
ENDNOTES   305
INDEX   327

PROLOGUE

It was 2021, and David Anthony was recalling a project he had been involved in a little over a decade earlier. In the summer of 2010, he and his wife and fellow archaeologist Dorcas Brown had joined a Ukrainian–American team excavating in the steppes south of Donetsk, in still-peaceful eastern Ukraine. From prehistoric burials they extracted sixty-six fragments of human bone. These they subjected to state-of-the-art tests, learning much about when the individuals concerned had lived and what they had eaten. The dig ended and Brown and Anthony returned to their usual place of work, Hartwick College in New York. The bone fragments had given up all their secrets, as far as they were concerned, and they put them in a drawer. 'That drawer closed and stayed closed for ten years,' said Anthony.

One day, out of the blue, he received a call from David Reich, a geneticist at Harvard University. Reich explained that his team had developed a method of extracting and reading DNA from ancient human bones, and he asked Anthony if he had any bones on which he could test the method. The drawer opened again, and in a way, so did prehistory. 'This is an amazing revolution,' Anthony said, of the way that conversation had changed his life. 'It's just incredible.' He paused, visibly moved, and went on. Archaeologists could now go into an ancient cemetery and determine the hair, skin and eye colour of the people who were buried there. They could discover how those people were related, not only to each other, but also to people lying in distant cemeteries, and whether they shared diseases. For the first time, they could confidently identify migrants in the archaeological record. 'It changes the whole ball game,' said Anthony.

The ball game he was referring to was the study of prehistoric languages, and one prehistoric language in particular – the one whose descendants are spoken by nearly half of humanity today. The one to which he dedicated his career. The one that ancient migrants carried far and wide.

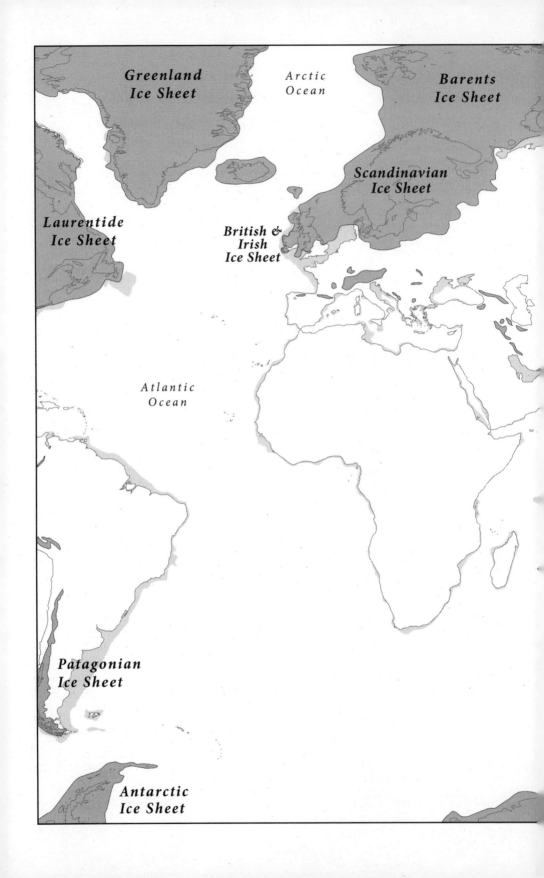

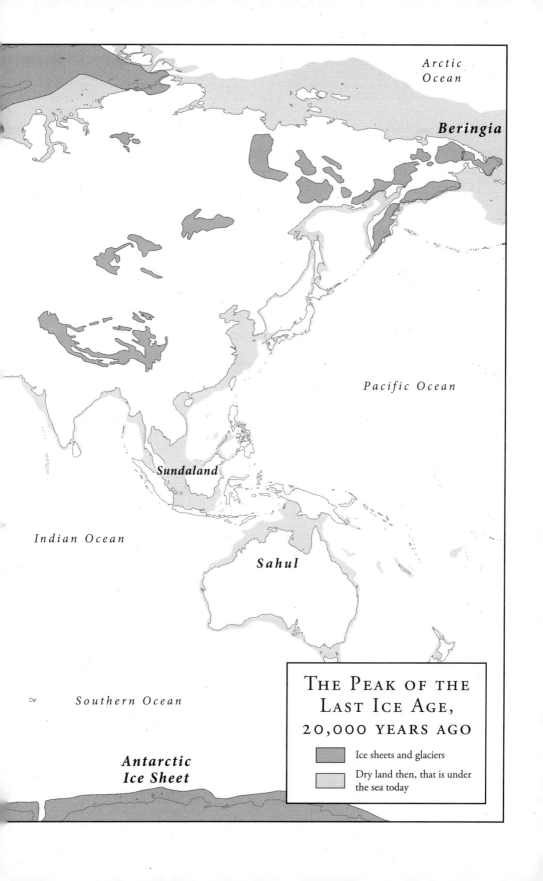

INTRODUCTION

Ariomania

The most powerful god in the ancient Indian pantheon was Father Sky. His name was *Dyauh pita*, literally 'sky father' in Sanskrit. For the Greeks the chief deity was *Zeus pater* – Zeus for short. The Romans deformed the *dy* sound of the original 'sky' word to give *Iuppiter*, or Jupiter. In Old Norse the *d* morphed into a *t* so that the Vikings recognised a thunderous god called *Tyr*, whose name in closely related Old English was *Tiu*. Tuesday is the day of the week that English-speakers dedicate to a god of weather and war.

Sanskrit, Greek, Latin, Norse and English are all descended from a much older language – Proto-Indo-European, from 'proto', meaning first, and 'Indo-European', the family to which those languages belong. The speakers of Proto-Indo-European, who might only have numbered a few dozen to begin with, lived between Europe and Asia in the region of the Black Sea. They too worshipped Father Sky. About five thousand years ago, their language exploded out of its Black Sea cradle, spreading east and west and fragmenting as it went. Within a thousand years, its offspring could be heard from Ireland to India. The Big Bang of the Indo-European languages is easily the most important event of the last five millennia, in the Old World. It took another three and a half thousand years and the invention of the ocean-going ship, but after 1492 some of those languages implanted themselves in the New World, and from there they expanded again.

Five thousand years is an eternity compared to a human life, and a heartbeat in the lifetime of humanity. Plenty of Indo-European languages and dialects have lived and died in that long/short interval, but more than four hundred are still spoken today. They span many of the languages of India, Pakistan, Afghanistan and Iran; the Slavic and Baltic tongues; Welsh, Irish and the other Celtic languages; English and her Germanic sisters; Greek, Armenian and Albanian; and the myriad offspring of Latin.

In Europe and some of its former colonies, people tend to trace themselves back culturally through the Jewish-Christian Middle Ages to the Romans and Greeks. Further east, many claim heritage in the Iranian languages and the prophet Zarathustra, or Sanskrit and the religions of ancient India. But go back further, beyond Ovid and the Greek myths, beyond the rambunctious gods of South Asia, and you hit a seam that connects east and west; a fibre stretched taut between them that thrums in all of us who speak Indo-European, though we may not be aware of it. This seam was forged in the heart of Eurasia before speech was written down – before history, that is, in mythological time. It teems with dragons and bears, wizards and warriors, formidable sky gods and wise, voluptuous queens. It's the realm of dreams and nightmares, of fantasies and fears. The Hindu scriptures, Homer's epic poems, *Beowulf* and *The Lord of the Rings* all tapped into it. All spoke to something very ancient in their audiences' psyche. All owe a linguistic debt to the first speakers of Indo-European. These people, for centuries an object of fascination and even delusion, have now stepped into the light. In the last decade, science has transformed our understanding of them. The language they spoke, its ancestors and its globe-trotting progeny, are the subject of this book.

INTRODUCTION

For all their swagger, Sanskrit and her sisters were a late development in the history of language, which is humanity's oldest tool. When *Homo sapiens* emerged in Africa three hundred thousand years ago, they were already equipped with it. Their vocal tracts could make all the sounds ours can. They had the cognitive capacity to map sound on to meaning, and to combine units of meaning, words, into sentences.

That's one view. Another is that language didn't evolve at all. It was invented in the deserts of south-eastern Africa around eighty thousand years ago, perhaps by a group of children left to their own devices, *Lord of the Flies*-style, and playing a game. In fact it may have been invented a number of times, in different parts of Africa.

The first human languages were spoken or signed or a combination of the two. They might have seemed quite rudimentary to us, lacking adjectives, say, or a fixed word order. But even with basic syntax their speakers (or signers) could transport their interlocutors beyond the here and now, beyond what was accessible to their senses. The ability to discuss the hypothetical is one of the most fundamental requirements of language. On that basis, it's not unreasonable to suppose that the first speakers could already tell stories.

Our ancestors were hunter-gatherers who roamed in small bands and chose their mates from outside the band. This would have brought new languages into the group, so that children grew up hearing more than one of them. Subconsciously selecting the most useful features of each, they refined the tool. By the time humans walked out of Africa sixty thousand years ago, they probably communicated mostly through speech and were capable of a more sophisticated level of expression.[1]

Sixty thousand years ago, the world was cold. It would grow colder. Glaciers eventually covered much of the western end of Eurasia, and where the ice stopped treeless tundra began. Humans followed their prey into warm-weather refuges. Storytelling, conjuring other worlds, might have become a survival skill. The glaciers began to recede around fourteen thousand years ago and the survivors moved out again. Within a couple of millennia, in the fecund half-moon of land that envelopes the Euphrates, Tigris and Lower Nile, their descendants began to invest more energy in tending plants and animals than in hunting them. A whole new world of food opened up to them.

On the eve of this farming revolution, the human population was an estimated ten million. Perhaps ten thousand languages were spoken, each one having one or at most two thousand speakers. Thanks to the new sources of energy that farming unleashed, communities grew and their languages with them. Dialects emerged, and in time some of them became languages in their own right. Families of languages erupted like supernovas. This period, the Neolithic or New Stone Age, was our linguistic heyday: the moment in the human story when more languages were spoken than at any other. At their peak there might have been fifteen thousand of them.

They weren't evenly distributed, because languages pool and split where humans do. An archipelago may contain as many dialects as it does islands. In the Caucasus, dubbed 'the mountain of tongues' by a tenth-century Arab geographer, linguists describe a phenomenon called vertical bilingualism, where people in higher villages know the languages of those living lower down, but the reverse is not true. Hotspots of linguistic diversity coincide with hotspots of biodiversity, because those regions can support a higher density of human groups speaking different languages, who don't need to stray. Melanesia and West Africa, glimpses of a once-global superabundance, still brim with languages today.

Languages mould themselves to the habitable landscape, then. But 'habitable' is, literally, a movable feast – one that is defined partly by the physical environment, partly by climate, partly by human adaptability. When the climate changed, people moved or adapted by other means, and their language, ever malleable,

INTRODUCTION

reflected their response. As Bantu farmers expanded south from central Africa, starting around five thousand years ago, their languages absorbed the clicks of the bushmen they met. As Alaska became more amenable to agriculture, Aleut words that once meant 'drop a fishing line' and 'distribute sea-catch' came to mean 'plant' and 'sow' respectively. There are limits to human adaptability, though. Greenland's Norse settlers vanished in the fifteenth century, along with their language, probably because a cooling climate pushed them beyond those limits.

From its Neolithic peak human linguistic diversity entered a long, slow decline. The downturn started with the formation of the first states, beginning five thousand years ago with Sumer in Mesopotamia (present-day Iraq). The languages in which those states chose to administer themselves grew, and within a thousand years the first ones with a million speakers had emerged. Some of them continued to expand at the expense of smaller ones. Many languages died, but rarely without a trace, since the survivors had rifled them for useful innovations.

Now, eight billion humans speak around seven thousand languages. Those languages fall into about a hundred and forty families, but most of us speak languages that belong to just five of them: Indo-European, Sino-Tibetan, Niger-Congo, Afro-Asiatic and Austronesian. Among those five, two behemoths stand out: Indo-European, whose major representative is English, and Sino-Tibetan, which includes Mandarin Chinese. Mandarin has more native speakers than English, but Indo-European has more native speakers than Sino-Tibetan. If you include second or subsequent language-speakers, Indo-European is by far the largest language family the world has ever known. That remains true if you measure it by geographical spread. Almost every second person on Earth speaks Indo-European.

Though ghostly whispers of lost languages weave through our modern speech, we'll never know what the vast majority sounded like, because they were never recorded. If you were to represent the three hundred millennia of *Homo sapiens*' existence as a twenty-four-hour clock, writing emerged at about thirty minutes to midnight. It was at that moment that history began (history, from Greek *historia* meaning 'knowledge' or 'inquiry', and later 'chronicle' or 'account'). Everything before that we call prehistory.

Like language, writing was conceived of independently a number of times. The oldest known writing system was invented by the Sumerians. It made use of ideograms, symbols that represent ideas pictorially. More abstract systems developed over time, in which symbols represented sounds. These fell into two main categories: syllabaries (in which each sign represents a syllable) and alphabets (in which each sign represents an individual speech sound). But there were also systems that were hybrids of the two, sometimes with a few ideograms thrown in.

As far as we know, the alphabet was only invented once, in or near Egypt. It was widely copied and modified each time. The important thing to understand about writing is that a given script can encode different languages, while a given language can be rendered in more than one script. Undeciphered ancient scripts, like those of Minoan Crete or the Indus Valley Civilisation, could therefore mask known languages or as yet undiscovered ones. Until someone deciphers them, we won't know.

If we relied solely on writing to discover language's past, this would be a very short book, since the Indo-European languages were spoken for millennia by people who never saw their names written down. Luckily we have other ways of interrogating that past. Because language is so malleable, because it has ceaselessly built upon itself, it is an archive of its own journey. When we talk about languages being 'born' or 'dying', we're defining a language as a package of communication tools that is unintelligible to users

INTRODUCTION

of other such packages. (Similarly, according to a standard definition in biology, species are distinct if they can't interbreed, or strictly speaking if their offspring can't interbreed.) That's a useful definition, but it's not the only possible one. It's no less true to say that all languages can be traced back to the first ones, just as all species can be traced back to the first living organisms. In that sense, all languages are equally old. Those we speak today are living fossils. Language is not only a tool; it's also a monument.

Humans have long known that the structure of a language and its relationship to others offer up a time capsule of the past. Herodotus, the Greek who is sometimes referred to admiringly as the father of history and sometimes less admiringly as the father of lies, grasped this intuitively in the fifth century BCE.[2] He wrote that when the inhabitants of the steppes north of the Black Sea wanted to trade with tribes in the Altai Mountains of Central Asia (the edge of the world disc, to him), they had to pass through seven interpreters and seven languages. He understood that languages could be both distinct and related.

Herodotus probably only spoke Greek. It took a polyglot, someone in whose brain an impressive number of languages jostled, to express that concept more clearly. One such person was Dante Alighieri, the man who single-handedly hastened the death of Latin by penning one of the most important works of medieval European literature in the Italian vernacular (in his compatriots' everyday language, that is). The impact of *The Divine Comedy* was such that, for a couple of centuries at least, Dante's Tuscan dialect became the standard literary language of western Europe.

At the turn of the fourteenth century, when Latin had already been shuttered into places of learning, Dante looked around him at Europe's linguistic landscape. Calling the languages by their word for 'yes', he distinguished the Germanic or *jo* languages from the Romance languages spoken to the south and west of them. Romance he further subdivided into the *langue d'oc* and the *langue d'oïl*, which divided at the River Loire, and the *sì* languages spoken in Italy and Iberia. Then he suggested that *oc, oïl* and *sì* were all descended from Latin.

His reasoning was that they shared many words, including those for god, love, sky, sea and earth. This jumps out at you even in the modern Romance languages, which have had more time to diverge. God, *deus* in Latin, gives *dieu* in French, *dio* in Italian and *dios* in Spanish. Love, *amare* in Latin, becomes *amour – amore – amar* in the same three languages; sky, *caelum*, becomes *ciel – cielo – cielo*; sea, *mare, mer – mare – mar*; earth, *terra, terre – terra – tierra*. In Occitan, the *langue d'oc* that is still spoken in pockets of France, Spain and Italy, the corresponding series is *dieu – amor – cèl – mar – tèrra*.[3]

Dante argued that any differences between these languages were the result of gradual change. He gave an analogy: 'Nor should what we say appear any more strange than to see a young person grown up, whom we do not see grow up: for what moves gradually is not at all recognised by us, and the longer something needs for its change to be recognised the more stable we think it is.'[4]

It's hard to express how heretical this idea was. Medieval Europeans had a different explanation for the languages they spoke: after Noah's ark ran aground on Mount Ararat, his descendants dared raise a tower to graze heaven. Affronted, God 'made a babble of the language of all the world'. All languages were the products of that punitive scrambling except one: Hebrew. This was the primordial language, the one that Noah had taken into the ark with him. Since none of these events had taken place more than a few thousand years earlier, there hadn't been time for the kind of gradual change that Dante espoused. He was barking at the moon, and so were those other madmen, his contemporaries who claimed that the Germanic languages shared a common ancestor too.

INTRODUCTION

> . . . learnèd philologists, who chase
> A panting syllable through time and space,
> Start it at home, and hunt it in the dark,
> To Gaul, to Greece, and into Noah's ark . . .[5]

Slowly, slowly, the evolutionary account gained on the biblical one. The idea was accepted that there were Italic, Germanic and Celtic *families* of languages. In the early eighteenth century, another polyglot, Gottfried Wilhelm Leibniz, claimed that these in turn shared an ancestor. By now Europeans were being sent out to manage far-flung corners of empires. Confronted by a dazzling array of peoples and tongues, some of them even learned those tongues, the better to administer the imperial subjects. In 1786, in a speech that has been quoted in countless linguistics textbooks since, a British judge and polyglot in Calcutta, Sir William 'Oriental' Jones, asserted that Sanskrit, Latin and Greek had 'sprung from some common source, which, perhaps, no longer exists'. He added that Germanic, Celtic and Iranian might have sprung from the same source.

Jones wasn't the first to express this idea, but his audience was finally ready to hear it. The suggestion that an archaic link existed between Europe and the Orient electrified the public imagination. In a world before aeroplanes and the internet, that loomed so much vaster and more mysterious than it does today, there was awe to be had in gazing upon Latin–Sanskrit word pairs like *domus-dam* (house or home), *deus-deva* (god), *mater-mata* (mother), *pater-pita* (father), *septem-sapta* (seven) and *rex-raja* (king). Or in comparing the first three numbers in German (*eins-zwei-drei*), Greek (*heis-duo-treis*) and Sanskrit (*ekas-dvau-trayas*). Of what ancient and fantastic encounters were these the fading echoes? 'Heard melodies are sweet, but those unheard are sweeter.'[6] Sanskrit studies exploded in the West.

Historical linguists, people who study how languages change over time, would eventually tease out twelve main branches of the Indo-European language family: Anatolian, Tocharian, Greek,

Armenian, Albanian, Italic, Celtic, Germanic, Slavic, Baltic, Indic and Iranic.[7] To write them in a list like that is to do violence to them, though, because each one hides not one but many worlds of human odyssey and thought.

Anatolian, a clutch of long-dead languages once spoken on the Turkish peninsula and including that of the great Hittite Empire, is considered the oldest branch, though it was one of the last to be recognised. Tocharian is another prodigal child, one no one expected to welcome into the fold, though the evidence that it belongs there is overwhelming. Also a dead language, or rather two, it was once spoken in the Silk Road trading posts of what is now north-west China, making it the family's easternmost branch.

The Indic and Iranic branches are considered so closely related, by most linguists, that they are usually combined into one: Indo-Iranian. By far the largest in terms of both geographical range and number of speakers, this branch also includes the lesser-known Nuristani languages that are spoken in remote valleys of the Hindu Kush. The Iranic branch comprises extinct Avestan, Zarathustra's mother tongue, and Sogdian, spoken by the merchants who plied the early medieval Silk Roads – the trade routes that connected Europe and China – but also Farsi (Modern Persian), Pashto and Kurdish. Their dead cousin Sanskrit, the language of the earliest Indian and Hindu scriptures, the Vedas, might be the most prolific of all the Indo-European languages. Among its many living offspring are Hindi, Urdu, Romani and the Sinhalese language spoken in Sri Lanka.

Greek, Albanian and Armenian are original enough that each one has a branch to itself. Their very orphan nature hints at a bevy of ghost languages, a clamour of long-dead relatives that might have travelled through time with them before expiring one by one and leaving only those three. The two surviving Baltic languages, Latvian and Lithuanian, have history entwined with the languages of the Slavs, judging by their closeness. But the Slavic tongues tell a tale of their own, falling out into western (including Polish and Czech), southern (such as Bulgarian and Slovenian) and eastern (notably Ukrainian and Russian) sub-branches.

At the other end of Europe, the Celtic languages divide north–south. Irish, Scots Gaelic and Manx took a different path from Welsh and Cornish, which show more affinity with Breton and extinct Gaulish across the Channel. Italic embraces Latin and its brood, from Portuguese to Romanian, but also Latin's dead sisters Umbrian and Oscan. Oscan was the language of the Sabines, whose women, legend has it, were raped by Rome's founder and his gang. Last but not least, Germanic spans the languages of Scandinavia along with English, Dutch and German, but also extinct Gothic, which was once spoken as far east as Crimea and southern Russia.

Most languages are themselves multiple. Take Modern English, which descends from Old English via Middle English. Linguists consider that on average it takes between five hundred and a thousand years for a language to become incomprehensible to its original speakers (who are, obviously, no longer around to look bemused). Modern English-speakers can understand the Middle English of Shakespeare, who wrote in the sixteenth century, but not the Old English of *Beowulf*, which was composed nine hundred years earlier. The first few lines of *Beowulf* are: *Hwæt. We Gardena in geardagum, þeodcyninga, þrym gefrunon, hu ða æþelingas ellen fremedon.*† American poet Stephen Mitchell translated them thus:

> Of the strength of the Spear-Danes in days gone by
> we have heard, and of their hero-kings:
> the prodigious deeds those princes performed![8]

---

† In Old English, the letter ash, *æ*, was pronounced similarly to the *a* in *cat*. The letters thorn and eth, *þ* and *ð* respectively, were both pronounced *th* and were used interchangeably.

(Though it may seem like it, the Indo-European domination of Europe was never complete. Finnish, Estonian and Hungarian belong to the Uralic language family, while Basque survives as a rare pearl of an anachronism: an island of something older in a sea of Indo-European. In the first millennium CE, conquering Visigoths tried to update the Basques, and every Visigoth king trumpeted *domuit Vascones*. The expression, in the Latin language that the initially Germanic-speaking Visigoths took up as they moved west, means 'he tamed the Basques'. He didn't.)

Logically, the parent of all the Indo-European languages in turn sprang from an ancestor that it shared with other language families, and you could keep going back and back until you reached the very first languages that humans spoke. But the deeper you go in time, the harder it is to know if languages resemble each other because they are related by common descent, or for some other reason – because they have borrowed from each other, say – and beyond about ten thousand years ago it becomes impossible to disentangle these effects. So while nobody denies that linguistic 'super-families' exist (one has been proposed that embraces Indo-European and Uralic), studying them is a fairly fringe pursuit. Most scholars have turned their attention to questions whose answers are within their reach.

No sooner had the link between Indic and European languages been demonstrated than people began to wonder where their common ancestor had been spoken. Locating the Indo-European homeland has been the Holy Grail for many intellectuals and many not-so intellectuals for two hundred and fifty years. Enlightenment thinkers rushed to claim India as the birthplace of the Indo-European languages, believing Sanskrit to be the most archaic of them. You could be forgiven for thinking that anywhere would do, for some of those champions of rational enquiry, as long as it wasn't biblical. The North Pole was a candidate, as was the lost city of Atlantis, vanished beneath the Black Sea. Practically every land bordering that sea has been on the table at one time or other.[9] Some of them still are.

Dangerous elisions were made in the early days, between the idea that the Indo-European parent or proto-language was spoken in one *place*, and the idea that it was spoken by one *people* who belonged to one *culture*. Nineteenth-century European archaeologists with nationalist leanings latched on to the word 'Aryan', the name by which ancient Indians and Iranians referred to themselves ('I gave the earth to the Aryan,' sings the god Indra in the oldest Indian text), and relocated this supposed *Urvolk* much closer to home.[10] The Nazis pushed that fantasy to its absurd and sinister extreme, claiming that the first speakers of Indo-European had blue eyes and blond hair, made pottery in a uniform style, and lived in northern Germany. A moment's thought reveals that such a one-to-one mapping of people, culture and language is an illusion. Indigenous Australians consider themselves ethnically similar but speak hundreds of different languages, while English is spoken by over a billion people claiming a bewildering number of ethnicities and cultures. Though most modern scholars believe that there was one Proto-Indo-European language, and that it was spoken by real people in a real place, they don't assume that those people were ethnically or culturally homogeneous.

There is a minority of scholars who question the very concept of a homeland. They point out that in the real world languages don't split neatly. Instead, dialects shade into each other. (This would have been obvious to anyone who lived before the eighteenth century. It became less so after national borders and languages were superimposed on dialect chains.) The definition of a language is hopelessly political – a language is a dialect with an army and a navy, in the witticism usually credited to linguist Max Weinreich – and languages don't only change by vertical evolution, but by horizontal borrowing too. In sum, the popular tree model of linguistic evolution, with its crisp branchings from a single, ultimate root, is an oversimplification that has led us disastrously astray. All that happened at the Enlightenment, these sceptics say, is that a nationalistic origin myth replaced the biblical one. The faster we run towards the mythical cradle, the faster it recedes, because it never existed. It's a mirage.

Most linguists find this view unduly pessimistic. They acknowledge that a tree model is a simplification, and that languages change through both horizontal and vertical processes. But they feel confident that they can distinguish these processes and walk them back through time. And they do believe that one can speak of the 'birth' of the Indo-European languages, and hence of their 'birthplace'. Aware that this mindset has led their discipline into some dark ideological cul-de-sacs in the past, their response is not to bury that past but to haul it up for all to see – a warning never to go to those places again.

The Indo-European language family enjoys the dubious honour of being the one on which historical linguists cut their teeth. Later, they took the skills they honed there and applied them to other language families. Indo-European is consequently the best documented and in many ways the best understood of all the world's language families, but it also drags the most outdated intellectual baggage behind it. It's like the star patient of a tail-coated nineteenth-century doctor, hauled out woozily for public display, underwear slipping off its shoulder, fêted and abused in equal measure.

How do you study a language that has been dead for thousands of years and was never written down? The short answer is with humility. A longer answer is with linguistics, archaeology and genetics. Historical linguistics probes the history that languages carry within themselves; archaeology tracks ideas and knowledge, the ingredients of culture; genetics tracks people.

INTRODUCTION

Though there is no one-to-one mapping of language, culture and genes, relationships do exist between these three, meaning that each one can be informative about the other two. Languages *broadly* reflect the cultures with which they are associated, because people tend to have more words for the things that matter to them, whether they be chisels or elves or salted, fermented sea herring (*surströmming* in Swedish). When people move, they carry their cultures and languages with them – at least for a while. Migration is considered a major if not the main motor of language change, because it drives a wedge between dialects and brings them into contact with different languages. On many continents, there is a correlation between prehistoric migration routes and the branching of linguistic family trees. Plenty of exceptions to these rules exist, if only because genes and languages are transmitted differently. (A person receives their genes from their parents, but they can receive their languages from a wider circle, even from books and apps, and they can lose them too.) Nevertheless, if you can track migration, the evolution of culture, and language, and cross-reference the three, you can start to tease out the human stories behind the relationships that were perceived dimly by Herodotus, and with greater acuity by Dante, Leibniz and Jones. You can reconstruct humanity's linguistic past, and even the long-dead languages themselves.

By the nineteenth century, the field of historical linguistics had been established on a more scientific footing. Using everything at their disposal, from ancient texts preserving dead languages to their living, spoken relatives, linguists systematically compared the formation of words and sentences across those languages. They described laws that predicted how a sound in one branch of the Indo-European family changed in another: how a *p* in Latin reliably became an *f* in English, or a *qu* became a *wh* – as in *pater-father* or *quod-what*. These particular sound transformations form part of a set known as Grimm's Law, after Jacob Grimm. When Jacob and his brother Wilhelm went hunting fairy tales in the German backwoods, it was partly to build a body of material in different dialects from which to reconstruct their Proto-Germanic ancestor. Among the happy by-products of their efforts were *Snow White* and *Little Red Riding Hood*.

The sound laws work because speech sounds drift over time, but the human vocal tract is not capable of an infinite variety of sound combinations. The directions in which individual sounds can drift are limited, and if one sound changes it tends to carry its neighbours with it. Guided by these laws, which restrict the possible ways languages can diverge from a common ancestor, linguists could start to identify inherited features shared by related languages, even if they sounded quite different in each. They could also differentiate between inherited and borrowed features, or loans. Loanwords could be highly informative in themselves, though, acting as a tracer dye of contacts between languages. It's partly thanks to loans that historical linguists were able to reconstruct the Roma people's thousand-year exodus from India. The Romani language is descended from Sanskrit, which is evident from the sound changes that separate the two, but Romani also picked up many loanwords during its epic trek west. Among those it assimilated in Persia were words for 'honey', 'pear' and 'donkey'. The Roma must have moved on from Persia by the time the Muslims conquered it in the seventh century CE, however, because Common Romani contains no Arabic loans.

> The winds lesser and greater
> cradled the little Gypsy
> and blew her far away into the world . . ."

Because languages preserve information without their speakers generally being aware of it, they offer up an uncensored view of the past. It may be an incomplete view, just as every historical account is incomplete, but when both language and history books are available they complement each other in interesting ways. The challenge, for linguists, is to extract their version intact, without being lured into the many traps that languages set for them. They had to learn not to be bamboozled by words that resemble each other by accident, for instance, or because they are the first words uttered by babies ('mama' is universal), or because they are onomatopoeic.

An example of onomatopoeia is the English word 'barbarian', which shares a root with *barbaras* in Sanskrit and *barbaros* in Ancient Greek. With apologies to Barbaras everywhere, the Sanskrit word, like the Greek one, refers to someone who stammers or speaks a foreign tongue. Both words, along with 'barbarian', were probably inspired by the sound people registered when they heard an unintelligible stream of speech: bar-bar or blah-blah or rhubarb. But *everyone* heard blah-blah when foreigners spoke. The Ancient Hebrew word *balal*, meaning 'babble', shares the same blah-blah sense as its English translation, and Hebrew is a language of the Afro-Asiatic family (Semitic branch). It would have been a mistake to conclude that Hebrew and English shared a common ancestor.

Gradually the linguists developed a sense of the relative age of linguistic features – certain sound combinations, say, or grammatical devices – based on whether they were present in older or younger branches of the family. This allowed them to attempt the delicate task of reconstructing long-dead proto-languages – the common ancestors of the twelve main branches and ultimately the mother of them all, Proto-Indo-European.[12] In recognition of the essential unknowability of these proto-languages, which predated writing, they adopted the convention of placing an asterisk before a reconstructed word, to indicate that it was hypothetical, that it had never been documented. *klewos*, for example, is thought to have meant 'fame' in Proto-Indo-European, or perhaps something closer to 'that which is heard' (that of which the poets sing). It was reconstructed from its descendants, which include Greek *kleos*, Old

Church Slavonic *sláva* and Sanskrit *shravas*, with the help of the sound laws.

It wasn't an exact science. The more branches of a family that contained features deemed to be related by inheritance, the stronger the case that those features were ancient; but people disagreed about how many branches were enough, and there were many disputes. Some of the most acrimonious were over the *meaning* of reconstructed words, since a word's meaning can shrink, stretch or sashay over time. The English word 'focus', referring to the point to which attention is drawn, comes from a Latin word meaning 'hearth'. The original sense of 'merry' was 'short', but at some point, probably because lessons or religious ceremonies that don't last long are reasons to be cheerful, it made a semantic leap. Words can moonlight too, taking on second shifts as new concepts are born and demand labels. English 'mouse', from Latin *mus*, can refer to a small rodent or a manual device for moving the cursor on your computer screen, but when the Romans used it they only meant one of those things. Much of the pleasure of etymology lies in charting the tortuous path that each word's meaning has travelled from its root to its modern usage.

It was as rare as hen's teeth that the linguists were vindicated, that an inscription or document was unearthed that proved a reconstructed word or sound had once been uttered, but it did happen. Ferdinand de Saussure, a Swiss linguist, postulated a vanished consonant called a laryngeal to explain the evolution of words that had once contained it, and a Hittite tablet later came to light that preserved such a sound in a symbol. Unfortunately, he didn't live to see it.

Historical linguists still work by comparing languages, though these days it's computers that lay family trees over groups of shared linguistic features, to find the best fit. But the comparative method can only tell you the *relative* age of languages; it can't tell you when, in chronological time, they were born, split or died. Loans can help with this, if they come date-stamped, so to speak, as in the case of the Persian words in Romani. Otherwise the only way to date key events in the life of a language is to turn to external, non-linguistic sources.

Histories are helpful. Roman chroniclers relate, for example, that when the future emperor Hadrian addressed the Senate around 100 CE, the senators mocked his Spanish accent (he was born in what is now the Spanish province of Seville). The fragmentation of Latin was underway, but Hadrian still spoke recognisable Latin rather than an early version of Spanish. By the time two of Charlemagne's grandsons swore a military pact in the ninth century CE, Romance had hived itself off from Latin. We know this because the two brothers, one of whom spoke Romance and the other German, made their oaths in each other's language for the benefit of their followers. The German-speaker, Ludwig, could have got by in Latin without notes, but Romance was a trickier prospect for him, and given how important it was to avoid a *malentendu*, he felt the need for a cribsheet. That sheet, known as the Oaths of Strasbourg, is the oldest surviving text in French, and indeed in any Romance language.

The *Rig Veda*, the oldest Indian text of any substantial length, is thought to have been written around 1400 BCE. The oldest Greek texts are roughly the same age. There are older Indo-European inscriptions – graffiti, epitaphs and other brief word strings – and the oldest of all are Hittite. These date to around 2000 BCE. As you go back beyond that date, beyond the point where the permanent trace of language flickers and blinks out, a new means of calibrating the trees is needed, an alternative way of divining and dating ancient people's dramas. This is where archaeology and genetics come in.

It was archaeologists of the nineteenth and early twentieth centuries whose discoveries of inscriptions and texts allowed linguists first to decipher various writing systems, then to methodically study the languages they encoded. But the discipline would go on to contribute in many more ways than that. The *science* of archaeology has come a long way in the last seventy years, and it's now entirely gripping. One archaeologist I listened to recently laughed in sheer delight as he described how dental plaque lifted from the teeth of shepherds who had lived thousands of years ago contained particles of charcoal from the smoke they had breathed in. An analysis of the particles revealed the species of tree they had burned.

The ratio of isotopes in bones and teeth speaks not only to what a person ate, but also to whether they grew up in the same place as their parents or were the child of immigrants, and whether they knew periods of hunger. (Isotopes, different forms of the same chemical element, occur in varying ratios in nature and hence in our food.) Bone scans show if they walked a lot, carried heavy loads or rode a horse. Such powerful tools had begun to yield hints that ancient people migrated even before geneticists could track the migrants themselves.

A seashell once worn as a bracelet can now be traced back along a chain of gift exchange to the beach on which it was picked up. A copper ingot salvaged from a shipwreck can be matched to the individual mine, half a continent away, from which the ore was extracted. The smelting of that ore might have left a swirl of lead dust deep in a peat bog. A scientist can drill into the bog, detect the swirl, and from it gauge the scale of the smelting operation.

Because peat accumulates so slowly, it acts like a carbon copy of the past. So does the organic detritus that sinks to the bottom of lakes. Pollen gets trapped in both, and by measuring the concentration and species of pollen in different layers you can reconstruct a sequence of prehistoric climate change, a time-lapse movie of rising and falling sea levels, advancing and retreating forests. Then you can ask, how do human sagas map on to that? Radiocarbon dating, at its most precise, allows the age of organic material, including bones, to be determined to within a human generation.

INTRODUCTION

'And I rose from the dark,' wrote the Irish poet Seamus Heaney, of an ancient bog burial.[13]

But even though archaeologists can extract whole libraries of information from fragments so tiny they can only be seen under a microscope, their knowledge too is partial. They can see that people moved, but not necessarily how many people, how fast or over what period of time. They can peer into the most intimate rituals of an individual's life, and death, but only guess at the beliefs that motivated them. There are nomadic peoples for whom no settlement has ever been found, and whole cities whose dead are missing (we'll encounter both kinds of absence in this book). In general, archaeology captures events rather than processes, but many of the forces that have shaped humans and their languages were slow and cumulative. So slow, at times, as to be imperceptible to the people upon whom they acted. It took another science to probe those processes.

Geneticists first learned to extract DNA from very old human remains about twenty years ago. That wasn't the first time genetic techniques had been applied to the study of prehistory, but until then people had looked for traces of past migrations in modern populations. A living person's genome (their full genetic complement) is a snapshot of their ancestry, all those forebears who contributed DNA to them, so this was a clever strategy. It's the strategy pursued by personal genomics companies such as 23andMe. But it is only reliable up to about ten generations back, because later contributions dilute earlier ones. It can tell you if your ancestor was brought to Algiers on an eighteenth-century slave ship, but not if another one mined salt in the Alps two thousand years earlier.

Reading the DNA of the long-dead was something that many people had thought would never be possible, because of the risk of contamination. Even brushing an ancient bone with your fingertips can introduce your DNA to it, and there were few bones in museums that hadn't been handled without gloves. (In the nineteenth century, archaeologists would even lick bones to gauge how old they were. The more one stuck to your tongue, supposedly, the older it was.) The breakthrough came in the early 2000s, when geneticists succeeded in identifying the distinctive profile of ancient DNA, the changes it had accumulated over time, and developed techniques for sorting it from the modern. Along with the scaling up of sequencing, which slashed the price and the time it took to read a genome, the ability to extract, sort and decipher ancient DNA constituted a sea change in the study of the unwritten past.

Since then, ancient DNA papers have been tumbling out. In 2023, the landmark was passed of ten thousand ancient genomes analysed, a hundred-fold increase over a decade earlier. It's not enough, by the geneticists' own admission; they need many more to fill the gaps in our knowledge of extinct peoples and to reconstruct those elusive processes. But the mass of detail they have already added to the portrait of prehistoric Eurasia is astonishing. Because they can track specific kinds of DNA that are inherited only through the male line (the Y chromosome) or only through the female line (mitochondrial DNA), they can distinguish the movements of men and women and map out mating networks. They have detected taboos on elites marrying beneath themselves and segregation between ethnic groups. They have picked up genetic signals of fostering, compassion for the disabled, human sacrifice, genocide and plague. They have determined the scale and in some cases the speed of migrations which archaeologists had only observed in freeze frames. Above all, they have confirmed beyond reasonable doubt the huge role that migration has played in the story of humanity and its languages.

Thanks to these advances, the study of the Indo-European languages has entered a thrilling new phase. It has become possible to triangulate in a way that it wasn't for most of the field's two centuries, and counting, of existence: to describe the events that turned dead languages into modern ones. Other disciplines have lent their weight. Mythologists piece together the stories that ancient people told by applying the comparative method to the

INTRODUCTION

building blocks of myths. In doing so they open a window on how those people understood the world. Ethnographers highlight possible parallels between modern and ancient societies. They tell us, for example, that herding societies are generally more violent than farming ones, because their wealth is mobile and hence more susceptible to theft. Computational biologists track the microbes that have evolved alongside us, including the ones that aid our digestion and those that make us sick. They are giving voice to what anthropologist James C. Scott has called the 'loudest silence' in the archaeological record: infectious disease. Germs, too, have played their part in shaping the languages we speak.

The origin of the Indo-European language family is one of the great outstanding problems of intellectual history, but reconstructing that family's past is a very peculiar type of exercise. The phenomenon that scholars are attempting to understand is ephemeral: the emanations of long-vanished brains that caused long-vanished eardrums to vibrate. They are acutely aware that the rules that guide them are, with the exception of the sound laws, rules of thumb and no more. If migration has driven language change, it hasn't been the whole story. The Scythians rode into Ukraine and India but left their language in neither.† The Romans got as far as Britain, but Latin stayed (mostly) in France. Whoever carried Celtic to Ireland caused barely a tremor in the Irish gene pool.

The reasons why people switch languages, or resist doing so, are and always have been complicated. Because of this, especially when it comes to the early, unchronicled part of the Indo-European story, nobody claims to have any hard-and-fast answers. At best what they are engaged in is a kind of ordering of scenarios by probability, or as one archaeologist put it, 'controlled speculation'.[14] There is a lot more agreement than there was thirty years ago, however, and

---

† Scythian is pronounced *si · thee · uhn*, with the emphasis on the first syllable.

detail is being added all the time. As the story moves towards the present, and historical sources become available to them, scholars become more confident in their conclusions.

The irony of the exercise is that linguists, archaeologists and geneticists are barbarians to each other. They don't speak the same language. Archaeologists think in terms of cultures, recurring patterns of objects that define a group's identity in some way. Cultures rise and fall, but genes flow on, albeit with dilutions and concentrations, so that geneticists have a different concept of identity. Languages have another dynamic again, changing through both descent and contact, but they are no less intimately tied up with who we are.

There's nothing intrinsically bad about this scientific Babel, because identity *is* multifarious, as each of us is intimately aware. In 2016 a man on a bus interrupted the conversation between a woman wearing a niqab and her young son to tell her to speak English in the United Kingdom. This earned him a reproof from another passenger, who pointed out that the pair were in Wales and had been speaking Welsh. Sixteen hundred years earlier, a Roman diplomat strolling about the camp of Attila the Hun, on the Lower Danube, heard someone address him in Greek and turned to see a long-haired man clad in fur. 'A barbarian who speaks Greek!' he exclaimed. The man explained that he was no barbarian, but that he had once been a Roman merchant before being captured and enslaved by the Huns. He had bought his freedom, remarried, and come to prefer their way of life. 'Now I fight Romans.'

Though linguists, archaeologists and geneticists regularly disagree about how to interpret the evidence, that too could be viewed as a strength. They each have access to two other bodies of data against which they can test their theories. When the triangulation is working well, they challenge each other's biases, keep each other intellectually honest. Each of the three disciplines, when applied to the Indo-European story, is tapping an elephant blindfold and calling it a crocodile, a python or a gnat. Together they come closer to divining a seven-tonne pachyderm.

Let's plunge into this Narnia, then, this newly written prehistory. Keeping us company will be the linguists, archaeologists and geneticists who first found their way through the metaphorical wardrobe (and who in some cases have only just got back; they're still dusting themselves off). First stop, in Chapter One, the Black Sea and the world around it after the ice melted. That's the world in which an ancestor of all the Indo-European languages was articulated for the first time. We don't know much about that ancestor, but we know quite a lot about the world in which it emerged – and that it was an insignificant player in a rich linguistic landscape. Chapter Two describes how that insignificant language became, not just significant, but extraordinary; how in a later incarnation it radiated with Bronze Age nomads who recognised no frontiers. That later incarnation, Proto-Indo-European, would give rise to all the Indo-European languages that are spoken today.[15]

The third chapter tackles the fraught question of Anatolian, the oldest daughter of Proto-Indo-European – or not. Thereafter we'll follow the Indo-European languages as they meander through the Old World, from prehistory through history towards the present, tracking their major branchings. Chapter Four recounts the Tocharian story, Chapter Five the languages of western and central Europe – Italic, Celtic and Germanic. Chapter Six describes the easterly expansion that gave rise to the Indo-Iranian branch, and Chapter Seven the Baltic and Slavic languages which, remarkably, may have participated in that expansion. Chapter Eight is dedicated to the idiosyncratic yet linked stories of Armenian, Albanian and Greek. The Conclusion, reasonably enough, takes stock. It stands in the present and looks back, then it turns round and squints forward, into the brume.

I

GENESIS

Lingua obscura

The justly named Golden Sands resort, north of Varna on Bulgaria's Black Sea coast, boasts warm, shallow waters where children can paddle while their parents look on tranquilly from the beach. The continental shelf is wide here, protruding around fifty kilometres (thirty miles). It is submerged today, but there have been times when the sea level was lower and the shelf was exposed. For most of the past two million years, in fact, the Black Sea was not a sea but a lake, a large fresh or brackish pond cut off from the Sea of Marmara, the Mediterranean and oceans beyond. Periodic warming caused the Mediterranean to rise and spill over the rocky sill of the Bosporus, injecting a mass of salty water into the lake and reconnecting it to the world ocean.

The lake was most recently cut off during the last ice age, when much of the world's water was locked up in glaciers. The glaciers melted, the oceans rose, and the moment when the Bosporus plug could no longer hold back the Mediterranean came, in one telling, between nine and ten thousand years ago. Water roared over that giant weir with the force of two hundred Niagara Falls, triggering a tsunami that surged through estuaries and lagoons and flooded an area the size of Ireland.

The manner of the reconnection, if not the fact of it, is debated. Some say that it happened gradually, as the Black Sea overflowed into the Caspian Sea, the Caspian Sea regurgitated the excess and the oscillation between them eventually subsided. Others say that the water level in the Black Sea rose ten metres as opposed to sixty. If it 'only' rose ten metres, the area of land flooded would have been smaller, the size of Luxembourg rather than Ireland. Still others suggest that, because that prodigious wall of water had to pass through the slender bottleneck of the Bosporus, it would have taken time for the sea levels to equalise. The Bosporus Valley might have roared at full spate for decades rather than months, a wondrous sight and sound in itself.

The two American geoscientists who proposed the deluge theory in 1997, William Ryan and Walter Pitman, speculated that the tales told by traumatised eyewitnesses might have been passed down orally over generations, until eventually they inspired the flood myths of the Bible and the Epic of Gilgamesh. 'He who saw the deep' are the first words of Gilgamesh's poem, written four thousand years ago in Mesopotamia, while Noah witnessed 'all the fountains of the great deep broken up'. There's no way of testing Ryan and Pitman's theory, as compelling as it is (and flood myths are not unusual). But perhaps the more profound impact of those events on humanity was that the Black Sea, always a valuable resource in itself, now became a conduit for other resources, including genes, technology and language.

By the time it was reconnected, it was roughly the shape and size that it is today. The Greek geographer Hecataeus of Miletus likened it to a Scythian bow, with the southern coast representing the string and the northern one the curved staff. The Greeks called it the 'inhospitable sea' (*Pontus Axeinus*), until they colonised its bountiful shores in the first millennium BCE and renamed it the 'hospitable sea' (*Pontus Euxinus*). It teemed with fish that had been pursued through the Bosporus Valley, now

Strait, by dolphins, seals and minke whales. It was probably Turkish mariners who, pursuing the fish, encountered its treacherous squalls and dubbed it 'black'.

To the north of the sea lay the steppe, known there as the Pontic steppe in a nod to the Greeks.[1] To the east lay the rugged peaks of the Caucasus, to the south the mountains and high plateaus of the Turkish peninsula or Anatolia, to the west the wooded hills of the Balkans and the Danube flood plain. Each was a world unto itself, but they met at the Black Sea, and whatever was exchanged between them could be ferried back deep into the interior via the great rivers that empty into it: not least the Don, Dnieper, Dniester and Danube. (Hold on to that recurring D, a linguistic tale to which we'll return.)

Ten thousand years ago, the Balkans were inhabited by the hunter-gatherers who had seen out the ice age in Europe. Another group of hunter-gatherers had moved west from the Caspian Sea as the world warmed, settling around the marshes and lagoons of the northern Black Sea coast and the rivers that feed them. At that time, a stretch of river south of modern Kyiv consisted of a series of rocky cataracts and lakes known as the Dnieper Rapids.[2] Archaeozoologists, people who study animal bones including those retrieved from old rubbish dumps, say there were catfish in those rapids the size of baby whales. The Eastern hunter-gatherers squatted the riverbanks, spears poised over the brooding megafish.

(The catfish of the Dnieper were up to two and a half metres or over eight feet in length, and three hundred kilogrammes – over six hundred pounds – in weight. Catfish approaching that size still swim in European rivers. They terrorise archaeologists diving for Roman relics in the murky River Rhône, whom they have been known to catch by the flippers, only letting go when they realise that archaeologists are too big to swallow. They are wels catfish, where *wels*, the common name of the species in German, shares a root with English 'whale'.)

To the south of the Black Sea, in the Fertile Crescent, the farming revolution was underway. 'Revolution' is a somewhat misleading term, in fact, since the set of practices that we call farming came together over a long period of time, in different places, through trial and error. The hunter-gatherers living at the western edge of the Iranian Plateau, in the Zagros Mountains, were probably the first to domesticate the goat (only the second animal to be domesticated after the dog, whose wolfish origins lie deep in the ice age). They likely grew wheat and barley too. To the west of them, in Anatolia and the Levant – modern Lebanon, Israel and Jordan – other hunter-gatherers began penning sheep and cultivating chickpeas, peas and lentils. In time the aurochs, a wild ox with long, curved horns, joined the domestic herd. The first farmers would have needed new words to describe these plants and animals, and the tools they invented to harness them. They would have acquired a vocabulary of agriculture.

Farming became a true revolution when its practitioners started expanding out of the Fertile Crescent. Throughout the twentieth century, archaeologists argued over whether the farmers themselves had migrated, or it was just their inventions that had travelled – whether other populations had simply embraced their ideas. Genetics showed that the farmers had moved, and on a massive scale. But theirs was in no way a conscious empire-building project; with each passing generation they simply needed more land to feed the growing number of mouths. It was colonisation by leapfrog: an advance guard travelling on foot identified a promising new site up to several hundred kilometres ahead, and others gradually settled the land in between. They took their languages with them.

Farmers from Anatolia entered Europe via two routes. One stream crossed the Bosporus, reaching the eastern Balkans by 6500 BCE and then following the Danube inland. Within a thousand years they were building villages in the Carpathian Basin – the depression, centred on modern Hungary, that is bound by the Carpathian Mountains and the Alps. A second stream island-hopped across the Aegean and along the northern Mediterranean seaboard by raft or boat (rowing, not sailing), then headed north from France's azure coast. The two streams met in the Paris Basin, the lowlands before the Atlantic, and there they mingled before fanning out again. By 4500 BCE, descendants of the Anatolian farmers were all over Europe, as far west as Ireland and as far east as Ukraine. These movements took place over many, many generations, but the distances covered are still extraordinary when you think that, apart from the sea crossings, they happened on foot. There was as yet no donkey or other docile pack animal, no horse that wasn't wild, no wheel and hence no wagon.

As the farmers advanced, the indigenous hunter-gatherers retreated. They were so few, and their footprint in the landscape so light, by comparison, that the immigrants might have had the impression that they were encroaching on virgin territory, at least to begin with. Some of the displaced hunter-gatherers headed for the Baltic Sea; others may have joined those skewering catfish on the Dnieper (who would certainly have spoken a language that was foreign to them). Still others stayed in their ancestral lands but sought refuge in the hills, or in the densest parts of forests: places that didn't lend themselves to cultivation, meaning the farmers passed on by.

Occasionally, in a forest clearing, a farmer and a hunter-gatherer must have come face to face. The encounter would have been a shock for both. Roughly forty thousand years had passed since their ancestors had parted ways during the exodus from Africa, enough time for them not only to behave and sound different but to look different too. The farmers were smaller, with dark hair and eyes and, probably, lighter skin. The hunter-gatherers had that now rare combination of dark hair and skin and blue eyes. They had no language in common, and they likely had different ideas on just about everything, from child-rearing to death to the spirit lives of animals. From what archaeologists can tell, such encounters did not typically end in violence. Sometimes the parties exchanged knowledge and objects. Sometimes they interbred. In time, many hunter-gatherers converted to the new economy, ensuring that some of their genes, perhaps even some of their beliefs and words, were passed on. But in general they couldn't compete. Their way of life and their languages were on a fast track to extinction.

From the Zagros Mountains, the Iranian farmers expanded east across the Iranian Plateau, in the direction of Afghanistan, Pakistan and India, and north towards the Caucasus. Not even the formidable Greater Caucasus range, with its 'gloomy, mysterious chasms, into which the mists crept down, billowing and writhing like serpents' deterred them, though they may have hugged the Caspian coast where the mountains tumble to the plain.[3] Soon farming settlements dotted the foothills of the North Caucasus. North of *those* hills lay the band of wetlands called the Kuma-Manych Depression.[4] This might have held the revolution up for a while, but before long it had penetrated the flatlands we call the steppe. And since you couldn't grow crops in the steppe – at least not in that part of it, where conditions were often too dry – the steppe-dwellers selected the one component of the farming package that worked for them: herding. The idea might have percolated in from the west as well, from the direction of the Carpathians. At first the steppe tribes kept only small herds, for the purposes of ritual sacrifice, and they continued to live off hunting and fishing. In time the herds became a source of food and textiles for them too. And as the herds grew, those people were forced to make occasional forays out of their valleys in search of fresh pasture. They strayed into the open steppe, but never far, and they always returned.

By 4500 BCE, the physical and genetic barriers that had divided Eurasian populations for tens of thousands of years had begun to come down, but a new divide had opened up. This one was cultural.

It separated herders from farmers, those whose wealth was mobile from those whose wealth was immobile. The two economic models bred two different mindsets: one that prized self-sufficiency and lived for the present, the other that valued collective decision-making and planned for the future. Both the Bible and the Qur'an recount how this clash of worldviews led to the first murder, that of the shepherd Abel by his farmer brother Cain, but the clash is much older than the Abrahamic scriptures.[5] In the Black Sea region it started more than six thousand years ago, when farmers and herders found themselves cheek by jowl at two steppe boundaries: one in eastern Europe, the other in the North Caucasus. That encounter marked the beginning of a dance of death that, for millennia to come, would bind the two in mutual hostility and dependence. Each grew and attained new heights of sophistication thanks to the other, but any malaise that affected one affected the other too, and climate change periodically rolled the dice. It was against this backdrop that the Indo-European languages were born.

Vladimir Slavchev noses his car between the nondescript buildings of an industrial estate and parks by a rusting gate. Beyond the gate a stretch of wasteland slopes down towards Varna Lake to the south. It's late November 2022. The trees are bare, but they still screen the city of Varna beneath us, to the left, and the Black Sea beyond it. Glancing over my left shoulder, I can just make out the bluffs running north-east along the Bulgarian coast, up towards Golden Sands and beyond.

Security cameras train their lenses on Slavchev as he unlocks the padlock on the gate. Never was such an unprepossessing parcel of earth under such heavy if discreet protection, but that's because it was here, exactly fifty years ago, that a man named Raycho Marinov was digging a ditch for a high-voltage electricity cable when he noticed that he had disturbed some metal and flint objects. He turned his finds over to the local archaeological museum and subsequent investigations revealed the remains of one of the most advanced societies ever to grace prehistoric Europe.

Here on the hill, within sight of their settlement on the shore of Lake Varna, a wealthy community of artisans and traders had buried their dead amid sumptuous grave goods. Besides exquisitely honed copper weapons and tools of flint and antler, they had interred thousands of gold objects including diadems, sceptres, bull-shaped figurines and astragals (sheep knuckle bones used as dice in the ancient world, though these ones were cast in gold). The cemetery at Varna was in use for just a couple of hundred years either side of 4500 BCE, but the gold extracted from it far exceeded that found at all other fifth-millennium sites in the world combined, including those in Mesopotamia and Egypt. And this at a time when vanishingly few human beings anywhere had set eyes on a metal object. Varna upended thinking about global prehistory. It forced the carving out of a new age from the late Neolithic: the Copper Age. It was an archaeological sensation.

Slavchev, who is an archaeologist, shows me the eastern sector where his team dug the previous summer. Pointing to a patch of darker soil in the wall of a trench, he tells me that it indicates an as yet unexcavated burial. He still doesn't know exactly how far the cemetery extends. His predecessor, Ivan Ivanov, excavated intensively for two decades, with labour provided by seventeen long-haul prisoners under the supervision of their guards (the best workers he'd ever had, he told Slavchev). But in 1991 Ivanov decided to put the project on hold, until such time as the newly democratic republic of Bulgaria had more resources to allocate to it and his successors could apply cutting-edge tools. The significance of the cemetery was already beyond doubt, and had been since his discovery in 1974 of the spectacular Grave 43. The sheer quantity of gold in this grave, including bracelets, rings, a sceptre, a penis sheath, even a gold-spangled hat, indicated that the man buried there had been a chief or priest. He had been around fifty at the time of his death. A reconstruction of his face, modelled on his skull, revealed a high-foreheaded, patrician-looking figure with an aquiline nose.

Ivanov died in 2001, and the site lay dormant for twenty years. Slavchev waited in turn for Bulgaria to get richer, only reviving the project in 2021. A stocky man with an easy smile and a wiry brown ponytail shot through with grey, he has the unflappable air of one who operates on millennial timescales. 'These people have lain in the ground for thousands of years,' he says. 'Who cares if it's me or someone else who digs them up?' The one resource he does have plenty of is labour (no longer Soviet-era lifers, but paid excavators), and he makes use of it as long as the weather permits. Having completed the season's dig at Varna, he had redeployed his team to another site about forty kilometres (twenty-five miles) inland, up in the hills. There, on a grassy, west-facing terrace above which rose stands of oak in all the splendid shades of autumn, they were unearthing the remains of a pottery workshop where people of the Hamangia culture had once decorated pots with bright-orange pointillist swirls.[6]

Around seven and a half thousand years ago, long before Varna had distinguished itself, the Hamangia had been among the first farmers to settle these parts. They brought with them material reminders of their Anatolian roots, in the form of the white or pink-orange *Spondylus* shells that their relatives had collected from the islands and coast of the Aegean (*Spondylus*, a scallop, doesn't grow in the Black Sea, preferring the warmer, saltier waters of the Mediterranean). These they turned into bracelets, belts and pendants, or beads which they sewed into cloth, before they exchanged them for other prestige goods. Their exchange networks stretched north across the Dobruja Plain to the Danube and its delta on the Black Sea.

In many respects the Hamangia were typical of European farming communities at that time. They lived in small settlements of wattle-and-daub houses with thatched roofs, whose walls they may also have decorated with colourful swirls, and they kept cows, goats, sheep and pigs. But there were a few ways in which they set themselves apart, Slavchev explains. They had sufficient understanding of chemical and physical processes to be able to produce sophisticated artistic effects with their ceramics. By regulating the oxygen supply to their kilns, they could obtain a black surface with a metallic lustre or an intense red or orange, to order. A Hamangia potter, possibly a woman since potters often were, left behind the first known depiction of thought in the history of art. Twelve centimetres (nearly five inches) tall, made of dark, burnished clay, the sculpture depicts a man sitting on a stool with his elbows on his knees and his head in his hands. The Romanian archaeologists who found it dubbed it 'The Thinker' (it has a pair, 'The Sitting Woman'). It is seven thousand years older than Rodin's statue of the same name.

Slavchev and I are standing by a trench-in-progress. My eye falls on a sack of bones at its edge, and, following my gaze, a cheerful, ruddy-faced man in the trench puts down his spade and plunges his hand in. Pulling out a specimen about twenty-five centimetres (ten inches) long, he lays it across my palm. Slavchev identifies it as a cow's thigh bone. 'They used them as ice skates,' he says, beaming. The climate in Copper Age Bulgaria was slightly warmer and wetter than it is now, but a body of water might still freeze over in winter, especially in these uplands. Slavchev says that prehistoric people strapped on cow bones at the slightest provocation; a frozen puddle would do.[7] I picture a woman skating beneath a leaden sky, wrists folded behind her back, dreaming of the statuette she'll sculpt that day.

The Hamangia culture lasted for a thousand years. Towards the end of its existence, its bearers experimented with placing copper ore in their kilns instead of fashioned clay. They may not have been the first to extract copper from its ore, by the process we call smelting, but they were probably the ones who bequeathed that valuable skill to the people of Varna. There are sites in Bulgaria, notably one called Durankulak on the coast, where layers of earth containing Hamangia artefacts lie beneath layers of earth containing Varna artefacts. This is how archaeologists know that the splendours of the latter grew out of the humbler foundations of the former. A thread connected the two, a continuity of ideas and traditions. Varna people continued to value the precious *Spondylus* shells of their distant ancestors.

By 4600 BCE, the inventiveness that budded with Hamangia had bloomed at Varna. The population had grown, in the region to the west of the Black Sea, and copper production had been scaled up accordingly. The ore was brought from half a dozen mines across the Balkans in the form of azurite and malachite. These minerals were powdered and mixed with ground charcoal before being roasted in kilns fanned by bellows. At eight hundred degrees Celsius (nearly fifteen hundred degrees Fahrenheit), the copper bled out in glistening beads that could be tapped and moulded into implements and ornaments.

An aura of magic must have hovered around the early smiths, who drew this gleaming marvel from blue-green rock. (It is even possible, as we'll see, that they were early models for Faust, the doomed man who prefers human to divine knowledge.) But in the Bulgarian heartland of the copper industry they formed just one element of a skilled community that included metallurgists, casters, miners and charcoal-burners – a level of specialisation, and of organisation, that Europe had never seen before. Around Varna Lake alone, eight settlements accommodated this community along with its support industries and dependants. Each one had up to eight hundred inhabitants, twice as many as a typical Hamangia village, and they lived in large, often two-storeyed timber houses that were laid out along streets according to a preconceived plan. Beyond the settlements, which were protected by wooden fences, lay cultivated fields, grazing herds and the dead. The Varna cemetery excavated by Ivanov turned out to be just one of several, albeit the richest discovered to date.

In time, the smiths added gold to their repertoire. It was panned out of the rivers that rose in the nearby Balkan Mountains, probably using fleeces, and brought to Varna as grains or nuggets. Ground up and suspended in an emulsion, it was painted on ceramics (the same technique was used with powdered graphite to produce stunning geometric designs in black and gold). Smelted to remove impurities, it was cast into jewellery and sceptres or other symbols of power. And besides gold and copper, there was a third commodity that was vitally important to Varna: salt. Farming communities used this mainly for conserving food, and some have argued that demand for it was so dependably high that it was salt rather than metals that made Varna rich. Up the River Provadia, which flowed into Lake Varna, a productive salt spring attracted its own walled settlement complete with shrine and cemetery. Here, in an industrial-scale operation, ceramic bowls were filled with brine and placed in giant ovens to accelerate the water's evaporation.

By 4500 BCE, the Copper Age societies of south-eastern Europe had reached the zenith of their wealth and influence. Their mastery of pyrotechnology, for which they must have developed a vocabulary, had put them in a league of their own. And what they produced, others wanted. Grave goods resembling those at Varna and Durankulak, including gold jewellery with some of the same motifs, have been unearthed from contemporary cemeteries on Georgia's coast and at Trabzon in north-eastern Türkiye. More astounding still, in a world before wheels, the Balkan miracle found its way deep into the steppe. Copper traced to one of the principal mines supplying Varna was adorning the bodies of the dead at Khvalynsk, an important ritual centre two thousand kilometres (twelve hundred miles) to the north-east, on the banks of the Volga. The copper ornaments in the Khvalynsk cemetery were more crudely made than those at Varna, suggesting that the copper was smelted in the Balkans, then transported in the form of small rods or ingots to its destination. There it was reheated (at this point lower temperatures sufficed), flattened into sheets using hammers of stone and antler, and turned into ornaments using chisels – beaver incisors set into bone handles.

How those rods or ingots were transported over such large distances is not clear. There were no wagons yet, but couriers could have carried small packages overland (the journey from Varna to Khvalynsk could have been completed in about a month on foot). Larger ones might have been loaded into dugout canoes. Though no remains of canoes have been definitively identified on the Black Sea's west coast, from this period, miniature models of them have, and archaeologists assume that some kind of vessel must have been in use there, for transport and fishing, since the bones of dolphins, seals and even whales have been found at Durankulak.

A large canoe six metres (twenty feet) in length could have carried up to four people and the equivalent of two men's weight in cargo. The copper could have travelled from the Black Sea to the Volga via the waterways of the Kuma-Manych Depression and then inland to Khvalynsk. Oxen, by now beasts of burden, could have dragged the heavy canoes upstream on the return journey. On the capricious Black Sea, however, seaborne trade would have been restricted to a short summer season, and even then such a vessel would not have strayed beyond shallow coastal waters. (Fisherfolk out of the Turkish port of Sinop have a saying: 'The Black Sea has only three safe harbours: July, August and Sinop.')

Who were the couriers – and what language did they speak? Who were those mysterious people who travelled so far, risking ambush by hostile tribes, not to mention the wild animals that skittered or lumbered into their path? Woolly mammoths and rhinoceroses had vanished with the last of the ice, but they still had bears, lynx, wild boar, aurochs and lions to contend with. Lion's teeth, pierced so that they could be worn as pendants, have been found in burials from this period around the Black Sea. The American archaeologist David Anthony, whom we met in the Prologue, has long argued that steppe societies dispatched their chiefly elite on this perilous mission.

Anthony figures among the most ardent and imaginative sleuths of the Indo-European languages of the last half-century, and the first to appreciate that ancient DNA would rewrite their story. But it's through the archaeological record that he detects the rise of this elite, and in particular through the steppe tribes' changing death rites. The hunter-fishers of the Dnieper Rapids had buried their dead in communal pits with not much more than a few deer or fish teeth to embellish their corpses. Once herding had infiltrated the steppe, burials remained communal, but certain individuals began to stand out in the mass graves by virtue of their eye-catching regalia, including caps and breastplates made of flattened boar's tusks, and belts of mother-of-pearl. Copper beads of Balkan provenance appeared, signifying status. Later on, steppe cemeteries shrank and the dead began to be buried singly, sometimes under small earthen mounds known as kurgans.[8]

By 4400 BCE similar kurgans had appeared in the Lower Danube Valley, very close to the northernmost copper-working settlements. They contained a great deal of copper, but also curious polished stone objects carved in the shape of horses' heads. These strange objects also crop up in settlements from the Balkans to Khvalynsk. Anthony interprets them as the heads of maces or clubs, the type of weapon that a chief might wield. Others see them as tools for polishing metal and infer that the steppe envoys were artisans. Whoever they were, it probably wasn't only men moving around the Black Sea at that time. Mitochondrial DNA, the kind that passes from mother to child, has been found in burials at Khvalynsk that originated in Balkan populations. And there are hints that women and children were moving in the other direction too.

One of the most intriguing such cases is that of a five-year-old girl who was buried with lavish grave goods in the Varna cemetery itself. Geneticists later retracted their conclusion that she was related to the steppe tribes, on the grounds that her DNA may have become contaminated. But despite the enduring question mark over her ancestry she remains the subject of intense speculation because of her unusual diet. In the farming societies of the Balkans, meat typically formed a smaller proportion of people's diet than it did on the steppe, and women ate less of it than men. Yet more than half the girl's diet consisted of meat, a higher proportion than characterised most of the men around her, including the chief in Grave 43. Could she have been the child of high-status immigrants, the daughter of a chieftain hailing from the steppe?

The individuals buried under the kurgans north of the Danube appear to have been pretty diverse genetically. Some carried ancestry from the steppe, some carried local farming or hunter-gatherer ancestry, others carried mixtures of all three. One way of interpreting this diversity is to posit that the first steppe emissaries found local imitators who acted as middlemen in the copper trade, or local wives with whom they had children who, once grown, took up that role. The coveted metal may have passed through a chain of intermediaries, each of whom travelled a relatively short distance to convey it. Whatever was given in exchange has not survived, but it may have been perishable. Among the commodities that have been proposed are animal skins, cured meats and steppe plants with medicinal or hallucinogenic properties. (It's possible that nothing was given, in which case 'exchange' becomes 'theft', but scholars think this unlikely because of the durability of the trade network through time, and the genetic and cultural mixing that accompanied it.)

The more tantalising question, from the point of view of this story, is what language the couriers and their suppliers negotiated in (and what language the little girl's death rites were pronounced in). The languages spoken around the Black Sea at the end of the last ice age are, sadly, beyond the reach of linguists' reconstructions. We can nevertheless be close to certain that, before they entered into their long-distance exchange activities, the steppe dandies with their mother-of-pearl belts and the gilded incumbent of Grave 43 had no language in common. It is also clear that the customers from the steppe must have lacked a lexicon of metalwork and smelting, at least to begin with, since they also lacked the technology.

In all of recorded history, you'd be hard-pressed to find a single example of human beings trading in high-value goods without an effective means of communication. Usually what has happened, in situations where the parties initially lack a common tongue, is that they have developed a lingua franca or shared language of commerce. The eponymous Lingua Franca, the supposed 'Language of the Franks', was spoken in Mediterranean ports from the Middle Ages until the nineteenth century.[9] (It was probably spoken earlier, before the fall of the Western Roman Empire, just not written down for a long time.) It developed out of Latin, but not the classical Latin of Livy or Tacitus, the 'vulgar' form spoken by ordinary, mostly illiterate people – the soldiers, sailors, colonists and enslaved people who frequented those ports. It was endlessly disparaged by the urban elite, precisely because it was a language of faraway places, of souks and brothels and cockfights. It was a chameleon, taking on the colours of Italian, Catalan, Occitan or any of the other emerging Romance languages, depending on where it was spoken and by whom. In the hundreds of Roman colonies of North Africa, it was strongly influenced by Arabic ('sugar', 'artichoke' and 'zero' are three of many Arabic words that entered English via the Lingua Franca).

It is likely that a lingua franca was also in use in the Black Sea region five thousand years earlier. We can only speculate as to what it sounded like, but some scholars have proposed that an ancestor of all the Indo-European languages *was* that lingua franca – that this ancestor gained an early foothold as a language of trade, eventually being adopted by many of the populations involved in that trade. Linguists are mostly unenthused by this idea. They point out that lingua francas are tightly tethered to the activity for which they were forged, tending to exist alongside their speakers' mother tongues without replacing them. Becoming a lingua franca is not, in itself, a recipe for world domination.

Nevertheless, many linguists do agree that an Indo-European ancestor was probably the mother tongue of one of the partners in that Copper Age trade network – the language of the couriers from Trabzon or Colchis on the Georgian coast, perhaps, or of those who hailed from the steppe via the Rivers Volga and Don. They think this mainly because, in order to have produced the degree of divergence that they see between all known branches of the Indo-European family, the common ancestor must already have been spoken by then. And since we know that the Black Sea network operated for hundreds of years, that women moved through it and that children were born of mixed parentage, there would have been time for generations to grow up who spoke more than one of the languages involved in the copper trade, in addition to the lingua franca. These bi- or plurilinguals, who as natural mediators might have become wealthy and powerful in their own right, would also have acted as conduits for the influence of one language on another. Through them, the Indo-European ancestor might have absorbed words, sounds, meanings and grammatical constructions from the other languages in the network (while donating its own to them). If that is what happened, then the Indo-European languages that we speak today contain echoes of the Pontic seaboard as it accosted traders' ears over six thousand years ago.

Around 4400 BCE, signs of strain began to appear among the farming societies of south-east Europe. A site called Tell Karanovo, two hundred and fifty kilometres (one hundred and fifty miles) south-west of Varna, was abandoned.[10] Starting with the first farmers to settle the Balkans more than two thousand years earlier, people had built and rebuilt on that site almost without interruption, until finally, in the late Copper Age, they vanished. Slavchev says that he never feels the past weigh on him so heavily as at Karanovo. You can stand in a trench there today and let your eye wander up twelve metres (forty feet) of compacted human debris: the trash of a civilisation that lasted longer than Christianity so far.

Hundreds of settlements followed Karanovo into oblivion, many of them apparently burned by their inhabitants before they left. Within two centuries Varna had stopped producing its glittering marvels and its cemetery had fallen into disuse. Europe wouldn't regain its social and technical heights for over a thousand years. The *Spondylus* shells that the farmers had treasured when they were still in the Near East, and exchanged like charms ever since, were now the opaque shibboleths of a vanished world. At about the same time, Balkan copper more or less disappeared from steppe graves, and its couriers disappeared from the Balkans. It is possible that the steppe tribes had become dependent on the metal, that its distribution in the form of rewards and tributes had become a vital means of keeping the peace among them, and when it dried up peace broke down. At any rate, there were no more sacrifices at Khvalynsk, and no more feasts. The shamans fell silent.

The balance of power was shifting in the Black Sea region, and the linguistic landscape with it. The Indo-European ancestor evolved and fragmented as its speakers' circumstances changed. And as it expired, its daughters entered the scene.

Alexey Nikitin hails from the village of Pochuyky, where steppe shades into forest-steppe south-west of modern Kyiv. His wife grew up across the street from him. Both played as children in the shadow of a vast earthen mound, the tomb of a Scythian warrior who died in the first millennium BCE. These days the mound is unimpressive; farmers have ploughed it nearly flat. But when Nikitin was five years old, he says, 'You could see it for miles.'

Pochuyky existed long before the Scythians arrived. One of the oldest continuously settled places in Ukraine, it lies in a corridor through which farmers from the west and herders from the east advanced and retreated over thousands of years. Nikitin, a palaeogeneticist, considers that his blood contains traces of all of them. He's the latest in a long line of migrants, and is used to moving on when things get tough. These days he works in the United States, at Michigan's Grand Valley State University, but the burial mound at Pochuyky still looms large in his imagination. He thinks deeply about the boundary that it marks, and how its meaning has changed over time.

Forest-steppe describes the band of sparse woodland that separates the forests of northern and western Eurasia from the treeless steppe, and it traverses modern Ukraine. When Varna was thriving, that ecological boundary coincided with a cultural one – the boundary between farmers and herders. The cultural divide was in turn reinforced by anatomical differences, with the gracile farmers and the taller, more robust herders still able to distinguish each other at fifty paces, and we can be fairly sure that different languages were spoken either side of it. Around 4200 BCE, however, the climate began to change, and the ecological boundary to slide.

Conditions grew cooler and drier around the Black Sea. Grassy steppe encroached on forest, expanding grazing opportunities for herders and making it easier for them to move around because rivers were fordable for longer. Life became tougher for farmers, on the other hand, since without rain their crops failed. One year of drought is survivable for the average farming community that sagely keeps grain in reserve. Two or three in a row can spell hardship, and they were now experiencing sustained periods of drought. It seems likely that, to add to their woes, their salt springs dried up. Control of this lucrative industry shifted to the populations living around the estuaries and lagoons further north. In the Dniester Delta, south of modern Odesa, the salt simply dries out on the flats; all you have to do is pick it up.

People began to leave Varna and the other Balkan settlements and move north, following the salt. In the lands beyond the Danube they would have encountered tribes living a more mobile lifestyle. To feed their families, some farmers may have converted to that lifestyle. (The evidence for this is that, in the Balkan uplands, no settlements were built for five hundred years following the collapse of the farming communities.) But they would have been novices at it: they didn't know how to manage large herds, or where the best pastures lay, or when those pastures were at their most succulent. They would have had to draw closer to the ancestral foe to survive. They might have started speaking a steppe language, no hardship for those who were already bilingual. But the power dynamic had switched: now they were the underdogs, the ones asking.

At some Copper Age Balkan sites, though not at Karanovo or Varna, there is evidence of extreme violence just before they were abandoned: massacres that spared no man, woman or child. This violence wasn't restricted to the Balkans. The archaeological record attests to growing tensions across Neolithic Europe at this time – manifesting, for example, in ever more robust fortifications around settlements – and these only intensified over the next few centuries. Knowing this changes the complexion of the strife in the Balkans. It could have been farmer-on-farmer, as climate change exacerbated social tensions within and between their communities, leading to scapegoating and feuds. David Anthony suspects that the steppe envoys may have delivered a final *coup de grâce*. Those peripatetic chieftains with their horse-head maces might have spied an opportunity to seize the means of production. It wouldn't have taken much. If they harried the farmers in their fields until the latter no longer dared leave their stockades, hunger would eventually have forced them to flee. Those who took their place would have spoken different languages.

A strange thing now happened. People who had previously inhabited small villages in the forests of Romania and western Ukraine began to move across the forest-steppe boundary, into the then sparsely populated area between modern Kyiv and Odesa.[11] Migrating east as far as the right or western bank of the Dnieper, they built settlements that were up to *twenty times* the size of their old ones. These astonishing megasites, some of which may have been home to ten thousand people, have puzzled archaeologists for more than a century. Some say they were simply farming villages that swelled as the population grew, others that they were refugee camps built to accommodate the farmers fleeing the calamity further south. Still others consider them a radical social experiment, proto-cities built along egalitarian lines.

Whatever they were, within a couple of centuries they were quite permeable to steppe influence. The farmers' ceramics were much finer than those of the herders, yet now the farmers started to import inferior pots from their neighbours in the flatlands. Archaeologists have been at a loss to explain why, but Nikitin thinks a possible answer is that they were exchanging, not pottery, but potters. By 4000 BCE, ancient DNA reveals that steppe women were moving into farming settlements, farming women were moving into steppe settlements, and these were no longer rare or sporadic events.

It's impossible to know if those women moved willingly or not, but whether they were wives or captives, they would have had to learn a new language – the dominant language in the new place

– while probably speaking to their children in their mother tongue (over the course of history, we know, captive women did just this). They may have passed on other skills, too, including knowledge related to food production. By the time the climate grew warmer again, around 3800 BCE, steppe herders in the contact zone were dabbling in crop cultivation. What had been a clear cultural and linguistic demarcation was becoming a continuum. And the boundary was softening further east as well, thanks to dramatic developments far to the south.

A few centuries after the appearance of the first megasites in Ukraine, Mesopotamian villages by the names of Uruk and Tell Brak also began to swell. These would become the first cities that prehistorians would recognise as such, because they were organised much like our modern cities – albeit extreme versions of them. They were dizzyingly hierarchical, with god-like rulers at the top, festooned in the trappings of power, and slaves at the unenviable bottom. As these southern cities grew, so did their appetite for raw materials. The kings of Uruk established a network of trading colonies, a mercantile dragnet to suck those materials in, and the Caucasus, so rich in timber, pasture and metal ores, marked its northern extremity. (As in the Balkans, people used fleeces to pan for gold in those mountain streams. It was in Georgia, in the ancient kingdom of Colchis, that the Greek mythological hero Jason found the golden fleece: '. . . he lifted the great fleece in his hands, and over his fair cheeks and forehead the sparkle of the wool threw a blush like flame'.[12])

By 3700 BCE, this other experiment in urbanism was making itself felt where the Caucasus meets the steppe. Farmers of Iranian ancestry had inhabited those foothills for hundreds of years, trading and interbreeding with the steppe-dwellers to the north. Now tombs on a truly monumental scale erupted from their midst. One kurgan was nearly as high as a four-storey house and the length of a football pitch in diameter. Beneath it lay a man, two women who had been sacrificed to accompany him to the afterlife, and a magnificent trove of treasure. Besides figurines in the form of a lion and a bull – Mesopotamian power symbols – the deceased took with him quantities of gold, silver, turquoise, lapis lazuli and carnelian, exotic luxuries from faraway lands. Under another kurgan, a man had been draped in an ankle-length coat made from the skins of two dozen *sousliks*, or ground squirrels.

These fur-coated princelings, or oligarchs, were the beneficiaries of that prehistoric gold rush.[13] Archaeologists disagree as to whether they were emissaries from Uruk itself, or local entrepreneurs, but having amassed fabulous wealth, they patrolled the limits of the Mesopotamian world oozing glamour and ruthlessness. They even took to their graves the cauldrons and meat hooks that had served them at banquets, and the gold straws through which they had sucked ceremonial beer. If they were colonisers, the piedmont seems to have been where their covetousness ran out. Their inner gaze was directed firmly south. Nevertheless, the new technologies that they flaunted at the steppe's edge attracted attention. They included the first bronzes, copper-arsenic alloys whose greater strength and flexibility translated into superior weapons, and possibly, the wagon.

Sometime between 4000 and 3500 BCE, the first wheeled vehicle trundled into the steppe, pulled by oxen. It's a measure of how transformative this technology was that evidence for it appears simultaneously east, west and south of the Black Sea, and archaeologists can't say exactly where it was invented. The steppe herders, by now seasoned middlemen in an international trade network, immediately saw what it could do for them. Couriers could transport heavier payloads: salt from estuary to megasite, ore from mine to forest-steppe (where the timber grew to fuel the smelting furnaces). Exactly how useful it was to begin with is disputed: the first wagons lacked a steerable front axis, meaning they could only move straight ahead. But roads weren't long in coming; archaeologists have traced a network of them connecting the crossing-places of rivers. And the hauliers, too, may have operated in bone-rattling relay.

Goods were certainly being traded in larger quantities from this time on, and they were travelling further too. Caucasian-style bronze daggers found their way, along with Baltic amber and Aegean coral, into settlements at the mouth of the Dniester. Fine pottery with the stamp of the Ukrainian megasites began moving across the Pontic steppe in bulk, as did silver. Mesopotamian cylinder seals, used for signing or sealing documents, have been found at sites on the north-east coast of that sea, that might have travelled up by boat from Trabzon.

By now the middlemen may have had access to longboats, powered by many oars, but on land the wagon-driver was king. East of Stavropol in the North Caucasus, a man was buried sitting up in one (his skeleton bore several healed fractures, an occupational hazard).[14] Along with the goods travelled knowledge and beliefs. The Caucasian oligarchs' monumental death mounds were soon being imitated, not only in the steppe adjacent to them but in the Dniester Valley far to the west. In some settlements along that river, kurgans abutted the flat cemeteries favoured by the more egalitarian farmers. Languages surely spread and mixed too. It's possible that the wagon-drivers spoke an Indo-European dialect, a daughter of the common ancestor, which they ferried back and forth across the steppe.

Around 3500 BCE, the megasites in turn were abandoned. Since their occupants left no trace of themselves, not a single bone or tooth (they may have cremated their dead, or left them out for the birds), the demise of those sites is as mysterious as their rise. Some suspect that they were casualties of the next iteration of the dance of death, rendered uninhabitable by the incessant threat of raids from the steppe. Others think that after seven hundred years, the experiment in decentralised city-dwelling was shelved, perhaps for reasons of internal politics. The survivors scattered, moving closer to steppe settlements or retreating back behind the forest-steppe boundary.

What followed was a period of broken borders, of fluidity and fusion across the western steppe. And out of that vortex of genes and ideas arose a revolutionary new culture. Permanent settlements vanished. Kurgan cemeteries stretched in long lines across the interior grasslands. The people who built them were the first herders to detach themselves from the river valleys and to adopt a fully nomadic lifestyle: the Yamnaya. They expanded rapidly, sweeping before them the intricate tapestry of cultures that had preceded and given rise to them, until they came up against the old enemy again – the farmers in their settled villages. In the west they overran the boundary that had once passed through Pochuyky. In the east they crossed the Volga. And then they kept on going, taking the mother of all living Indo-European languages with them.

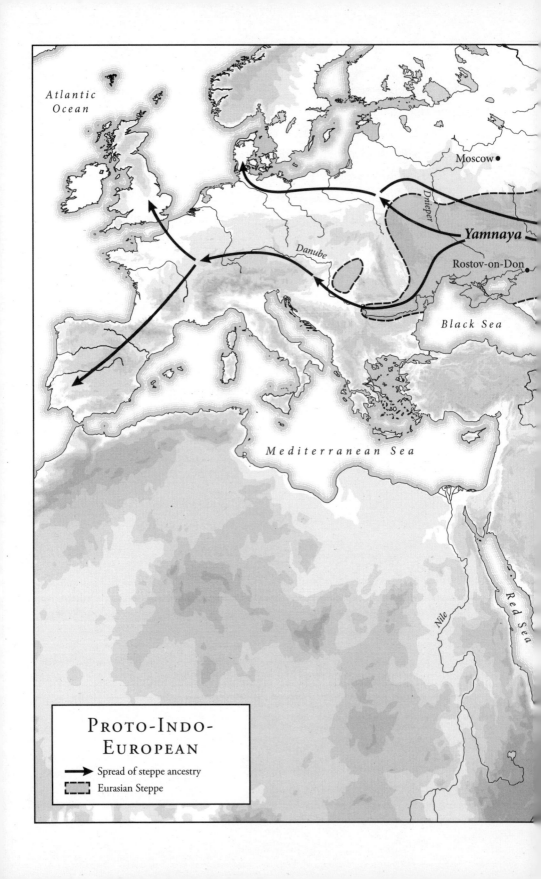

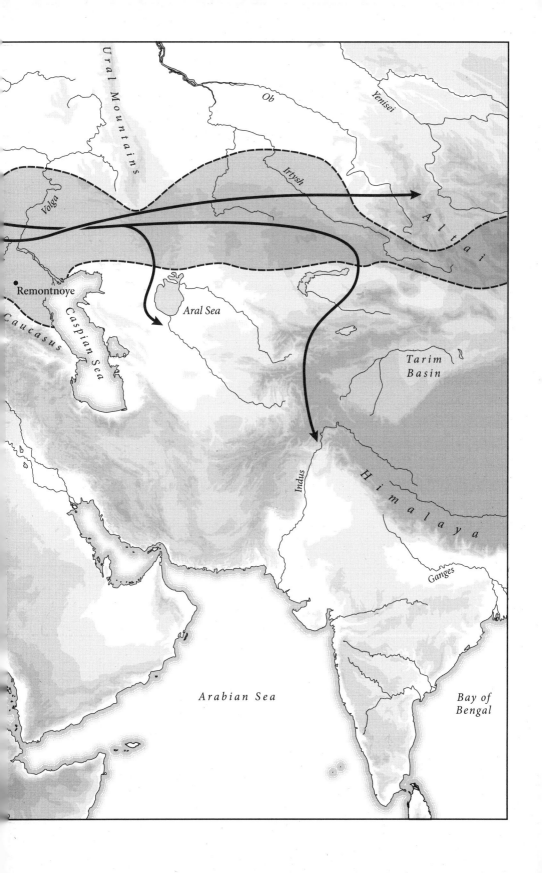

2

Sacred Spring

Proto-Indo-European

Few people realise how far the steppe extends into Europe. About half of Ukraine is steppe (*U kraina* means 'at the border') and it runs right down to the northern shore of the Black Sea, leaping over the Isthmus of Perekop into Crimea. 'When I am dead, bury me / In my beloved Ukraine, / My tomb upon a grave mound high / Amid the spreading plain,' wrote the Ukrainian poet Taras Shevchenko.[1]

A lick of steppe curls around the north-west shoulder of the Black Sea, reaching down through Romania to stroke the Danube Delta. To the west, it forms a gleeful and very emphatic full stop in the Great Hungarian Plain, right inside the Carpathian Mountains that present their backs so defensively to the east. From there it runs eight thousand kilometres (five thousand miles) back the other way, forming a belt across the middle of Eurasia. To the north of this belt are forests. To the south, between the Black and Caspian Seas, lie the snow-capped Caucasus Mountains. The nature of that southern frontier changes as the deserts of Central Asia loom, but the steppe sprints on until it reaches Manchuria. A little past the midway point rise the Altai Mountains, marking the junction of four modern nations: Russia, China, Mongolia and Kazakhstan.

Beyond the Altai lies the eastern steppe, which is higher, colder, drier and altogether less hospitable than its western counterpart. Not that the western steppe is hospitable. Temperature swings can be extreme, and it is dominated by grasslands for as far as the eye can see. Threading through these infinite prairies, irrigating them, are rivers and streams that run, as a rule, north–south. In parts of Ukraine and Russia where the fertile *chernozem* or 'black earth' dominates, the prairies have become wheatfields. East of the Don, rainfall is unpredictable and growing crops is out of the question. There the soil is 'salty' *solonetz* and you're more likely to encounter shrub-steppe, or desert-steppe. Before the dew burns off in the morning, or after a shower, it's the smell that strikes you: the bitter, herbaceous aroma of *Artemisia* or wormwood. 'You can quickly get drunk on that smell,' says archaeologist Natalia Shishlina. At all other times it's the sky, and the unbounded space. At night, the stars touch the Earth.

Shishlina and I are lifting our knees high as we tramp through the grass. Ahead of us Idris Idrisov, a geographer, is doing the same. It's July 2023, about nine in the morning, and the temperature is already pushing thirty Celsius (the high eighties Fahrenheit) in the southern Russian steppe. Usually by this time of year the vegetation has withered to scrub and the ochre soil is visible, but it rained so much in the last few months that the land is still green. With every step we take we put up a cloud of cicadas. Things change quickly here, Shishlina tells me. For two weeks every spring it's a carpet of tulips. In autumn, after rain, waves of mushrooms wash over the steppe. Local people take what it gives them when it gives it, because they know that the next minute it will be gone.

Shishlina works at the State Historical Museum in Moscow. She has two passions: ballet and the steppe – specifically the steppe about five thousand years ago, in the Bronze Age (the age of heroes, as she calls it). Every summer since the late 1980s, she and her team have taken the bus twenty hours south to excavate in that part of it that stretches between the Rivers Don and Volga. Here the landscape rolls gently from watershed to watershed, and prehistoric nomads left thousands of kurgans on the elevations. For years, though, Shishlina's team failed to find any of their settlements. To locate the campsites that she knew must exist, however faint a trace they may have left, she had to think like a nomad. Oddly enough, it was the collapse of the Soviet Union that helped her to do so.

In the early 1990s, almost nothing worked in Russia. Planes were grounded, factories slumbered, people did what they had to do to survive. Shishlina, who was excavating in Kalmykia at the time – a republic enclosed by the Russian steppe and bordering the Caspian Sea – realised that she had become a bystander to an unique historical experiment. Deprived of modern conveniences, especially petrol, the Kalmyks had reverted to their traditional lifestyle. Mounted on horseback rather than their habitual motorbikes, they led their herds from winter to summer pastures and back again, in a continuous round. Their herds had shrunk and the steppe had been restored to something close to its prehistoric state, minus the wild horses, aurochs and pale, bulbous-nosed saiga antelope that had grazed there in earlier times.

Shishlina interviewed Kalmyk shepherds about how they selected a campsite and searched in places that met their criteria. At the end of each field season, she got back on the bus empty-handed. Then Idrisov joined the team, from the Institute of Geology in Makhachkala, the capital of the neighbouring Russian republic of Dagestan. Shishlina says that everyone laughed at him at first, because he was always getting lost, but they soon learned to appreciate him for his ability to read the landscape. Idrisov told them that they had been looking in the wrong places, or rather in the right places but at the wrong elevation. The Yamnaya might have buried their dead on the ridges, for all to see, but they probably built their camps lower down, maybe two-thirds of the way up the slope, so that they were hidden from passers-by and sheltered from the elements.

He led the archaeologists on walking surveys, identifying promising valleys and then buried soils within those valleys that contained animal bones and pottery fragments, known to archaeologists as sherds. They dug experimental pits and finally, in the early 2010s, excavated their first campsite. There wasn't much to it: a few sherds, some metal objects. But others followed. In time they shifted their focus to Rostov Oblast, a little to the north and west of Kalmykia. The day I joined Idrisov and Shishlina on their survey, near the village of Remontnoye, Idrisov pulled a couple of sherds out of a low earthen cliff.[2] Shishlina said that they were probably Sarmatian, dating to the first centuries CE, but that she would bring the team back to investigate further, because later nomads often camped over the stopping places of earlier ones.

With the evidence furnished by campsites and burials, the archaeologists began to assemble a more complete picture of Yamnaya life. The herders who preceded them on the steppe had stayed close to their ancestral river valleys. They may have led their animals out in search of fresh pasture, when the grass at home had been grazed down, but their valley settlements remained occupied all year round. The Yamnaya took transhumance to the next level, living out of their wagons or tents and moving with the seasons. In Russian, Shishlina explains, the same term designates tumbleweed, the rootless plant that scatters its seeds as it rolls, and the nomads of the steppe: *perekati-pole*. The Yamnaya were the first *perekati-pole* and the model for all those who came after them, including Scythians, Sarmatians and Mongols.

Their way of life probably came together over time, around the middle of the fourth millennium BCE, because it could only have worked with certain types of tools and knowledge and a great deal of planning. Their wagons were pulled by oxen. In summer they were limited by the availability of water sources, in winter by snow cover (sheep and goats can't graze in more than thirty centimetres or a foot of snow), and they plotted their circuit accordingly. They did not yet have long-haired sheep that provided wool thread of the kind we are familiar with, but they may have plucked their sheep's short under-wool to make felt-like textiles that were suitable for tents. Skins and furs provided their clothing, and they burned manure for warmth.

Archaeologists' studies of microscopic traces of food trapped in dental plaque have revealed that milk was consumed rarely if at all on the steppe before the Yamnaya, but that it was central to their diet. To begin with it was exclusively sheep's milk; later they diversified into that of goats, horses and possibly cows. They ate these animals' meat, supplementing their protein-rich diet with wild plants such as barley and sorrel, sources of carbohydrate and vitamins that Shishlina suspects they may have replaced with dried tulip bulbs and mushrooms in winter.[3] *Ephedra,* chicory, wormwood and cannabis were all steppe plants whose medicinal or psychoactive properties they knew, and it has even been suggested that they domesticated cannabis. They traded with others for certain raw materials that weren't to be found in the steppe, such as honey and copper ore, but otherwise they were self-sufficient, smelting copper to make bronze and forging their own weapons and tools. Everything they had was either expendable or portable, even their altars, if those Russian scholars are correct who surmise that this was the purpose of the decorated bone pins found in some Yamnaya graves.

Once they had struck out into those oceans of grass they never looked back. They developed a taste for freedom and space that danger no longer cancelled out. The evidence is those long lines of kurgans, some of which were crowned with carved stone stelae. Each mound was raised initially over one individual, who had been placed in a chamber, often on his back with his knees raised. Typically, the deceased was laid on a mat woven from steppe plants, his head resting on a pillow stuffed with aromatic herbs, and he was sprinkled with red ochre. Compared to the Caucasian oligarchs whom they may have imitated, the Yamnaya went to the afterworld with modest grave goods: a few pots and astragals, some hair rings twisted out of silver wire, wagon parts and animal bones. A bronze knife might be placed under the head of an adult, but in general tools and weapons were rare.

As economies go, theirs was revolutionary, because it allowed them to unlock the vast energy reserves of the Eurasian steppe. Humans had hunted the wild animals which grazed those prairies for millennia, but the Yamnaya cast a larger net, with a finer mesh. They fixed that deep well of nutrients more efficiently than anyone had before, by converting it into sheep, goats and cattle, and thence into food, fuel and textiles. Their bones and teeth testify to this. They grew significantly taller and stronger than their ancestors. Some of them were remarkably long-lived. Archaeologists have retrieved sexagenarians, even septuagenarians. And very soon, there were a great many more of them: a veritable population explosion. If the herders of the Yamnaya culture were responsible for the radiation of the early Indo-European languages – and as we'll see, there are persuasive arguments that they were – you would have to say that that radiation was powered by the steppe.

> Dead grasses parched by heat. The steppeland, seared,
> Runs on and merges with the sky's pale reaches.
> Here is a horse's sun-bleached skull, and here
> An idol with its flat, stone features.[4]

That summer of 2023, while war raged across the Don and a mercenary named Yevgeny Prigozhin led a rebellion out of Rostov, a few hundred kilometres to the west, I watched Shishlina's team excavate their umpteenth kurgan. A dust storm forced them to down tools, briefly, but when they resumed work they discovered the remains of a Yamnaya man who had been laid in the ground around 2700 BCE. Assuming that enough of his DNA had survived intact, it would join the growing library of Yamnaya samples – a library that has enabled geneticists to track those probable speakers of Proto-Indo-European from their origins to their maximum expansion, and added telling detail to the portrait that archaeologists had already drawn of them.

Yamnaya is the name of an archaeological culture. It means 'pit grave' in Russian, and it is by its burial rite that the culture is defined. The bearers of that culture varied their style of pottery and metalwork, and their diet, but to start with, at least, their treatment of the dead never wavered. It was so consistent that some have suggested they were bound by a strict ideological or religious code. They were extremely homogeneous genetically too, in the beginning. What is most striking about the earliest Yamnaya males to be found to date is that they carried a very narrow cluster of Y chromosomes. Later on, after the population had grown, other Y chromosomes entered the mix, but the first of their kind may have been closely related to each other on their father's side.

One way to interpret the genetic evidence is to think of the Yamnaya as a single clan or brotherhood who distinguished themselves by their burial rite. They may have left a larger group, or been expelled from it, and having moved out of their ancestral valley become increasingly nomadic until they vanished into the grasslands for good. If that is who they were, it prompts an extraordinary reflection: fewer than a hundred people may have spoken the dialect that gave rise to all extant Indo-European languages. Scientists even claim to have homed in on the place from which they departed, since the oldest Yamnaya sites, which have yielded the oldest Yamnaya genomes, are located between the lower reaches of the Rivers Dnieper and Don, in the embattled east of Ukraine.

The evidence can be interpreted differently, however. A kurgan would have taken weeks or months to build, so it might have been an exceptional privilege. We already know that not everybody was buried this way, since in many parts of the Yamnaya range adult male kurgan burials greatly outnumber those of women and children. The Yamnaya may have been the elite of a larger population that doesn't show up in the archaeological record because, having built the earthen mounds of their overlords, they themselves received much simpler burials. If this alternative interpretation is the correct one, then the population speaking that ancestral dialect may have been larger, in the hundreds or thousands. The elite may have maintained its genetic 'purity' by marrying within itself, the men selecting wives from the same few related clans, as elites have often done.

The study of ancient DNA has restored flesh to arid bones, to the extent that you sometimes feel as if you can reach out and touch the nomads they belonged to. The Yamnaya were typically brown-haired and brown-eyed, geneticists say, with a complexion somewhere between fair and dark. They were extremely tall for their time, over a metre eighty or six foot in the case of some men. The women, at least those who received kurgan burials, were tall too. Their height may have been partly attributable to their protein-rich diet, but height is mostly under genetic control and it looks as if the population had captured a mutation that drove it upwards. On the other hand, there is no evidence that they were biologically disposed to digest milk, which is surprising given how important dairy was in their diet.

Inability to digest lactose, the sugar in milk, is the default state of human adults. Babies can do it, but the enzyme that breaks the lactose down is usually switched off by adolescence, such that the majority of the world's adults suffer cramps and other digestive symptoms if they drink more than a little milk. This made evolutionary sense, in our dairy-free hunter-gatherer past, because it meant that babies weren't competing with others in the group for their mothers' milk. But with the rise of dairying, as milk became plentiful, mutations that kept the enzyme switched on in adulthood became widespread. (In Europe, these mutations are more common in the north than in the south, reflecting the paths taken by the Yamnaya's descendants; this is the biological underpinning of the butter/olive oil divide remarked upon by the Romans.) But it looks as if those mutations started spreading only *after* the Yamnaya. That in turn suggests that the Yamnaya consumed their milk much as lactose intolerant southern Europeans do today, in fermented form. During fermentation, the process that produces yoghurt, cheese and kefir, bacteria in the milk break down the lactose so that your gut doesn't have to.

Having hit upon the winning formula of nomadic pastoralism, around 3300 BCE, the Yamnaya's numbers swelled and they began to spread through the steppe north of the Black and Caspian Seas (the Pontic-Caspian steppe). Within three hundred years they had pushed as far east as the Ural Mountains and as far west as the Carpathians. Perhaps the most contentious question, as regards this radiation, is whether it was propelled by the horse. Archaeologists are fairly certain that the Yamnaya hunted horses, but they are divided over whether they rode them – and whether this was how the early Indo-European languages expanded so far, so fast.

Janet Jones, a neuroscientist who studies horses and their riders, points out what a miracle of locomotion is the horse-human pairing. The horse is a prey animal, prone to becoming 'hysterical in a heartbeat'; the human is a predator. The average horse weighs eight times the average human, meaning that not even the strongest of us can make one of them do what it doesn't want to do. Yet together, with the right training, a horse and its rider can cover huge distances, leap blind over obstacles and attain speeds that weren't beaten by the motor car until the twentieth century: 'We run as one at speeds of up to forty-four miles [nearly seventy kilometres] per hour, the fastest any land mammal carrying a rider can achieve,' writes Jones.

But the process of domestication that culminated in that miracle was long and slow, and pinpointing its beginnings isn't easy. We know that east of the Urals, in what is now Kazakhstan, people belonging to a culture called Botai were already raising horses by 3500 BCE. They penned them, ate them and used their manure as roof insulation. They may also have ridden them and drunk fermented mare's milk. The Botai *only* kept horses (and dogs), and for a long time they were credited with domesticating them, but recent studies of ancient horse DNA suggest that the story may be more complicated.

While the Botai were selecting horses for docility and strong backs, and breeding from animals that possessed these traits, other people were doing the same to the west of the Urals. Starting from different wild stocks, both groups were honing a beast that in combination with a human would usher in Jones's miracle of locomotion. But only one of those animals would go on to replace almost all of the world's wild horses, and that was the one from the west of the Urals (*Equus caballus*). The last descendants of the Botai's line are the Przewalski's horses (*Equus przewalskii*) that still roam the Mongolian steppe and were once thought to be wild, but are in fact feral (having once been domesticated, that is, they have reverted to the wild state).

The people responsible for domesticating the western line, though the process may have begun before them and certainly continued after them, were probably the Yamnaya. Their equine stock only began to spill out of its steppe heartland several centuries after they had vanished from the archaeological record (I'll tell the story of the spillage in Chapter Six). But already in the Yamnaya's day, there were horses in that heartland that belonged to an almost identical line. They must have been tame enough to milk, because we know that at least some Yamnaya consumed mare's milk. The sixty-four-thousand-dollar question is: were they also tame enough to ride?

The reason this question is difficult to answer is that horse-riding is hard to detect in the archaeological record. The first horseshoes were probably woven out of plants that have long since perished. Early riding tack was made from leather or rope that has rotted too, and the first riders used very little of it anyway; saddles and stirrups were later inventions.[5] Scientists have looked for signs of wear on ancient horses' teeth, caused by friction from a bit, but this can be indistinguishable from natural wear, so they have tried looking for evidence of riding in human remains instead.

One recent study diagnosed 'horsemanship syndrome', a constellation of signs including compression of the lower vertebrae and changes to the hip joints, in five out of two hundred Yamnaya individuals examined. But the finding is controversial, with critics claiming that the changes could have been brought about by any activity that causes repetitive mechanical stress (even the authors admitted that basket-weaving or barrel-making could have caused them). Shishlina is one of those who remain unconvinced. The physical anthropologists she works with, who have pored over the bones that she has excavated, conclude that the Yamnaya spent a lot of their lives walking beside wagons, sometimes carrying heavy loads. Shishlina doesn't rule out the possibility that the Yamnaya followed wild horses to their winter pastures. She thinks that they may have exploited the '*tebenevka* effect', whereby horses break up deeper snow than cows do, and once they have done so the cows will graze. But she does not think that they rode them.

David Anthony takes the opposite view. He traces the beginnings of horse domestication back to Khvalynsk, fifteen hundred years *before* the Yamnaya. There shamans were sacrificing horses along with sheep and cows, which suggests to him that they already thought of horses as domesticates. Mentally, that is, they placed all three animals in the same category. The process continued with the steppe envoys who crossed the Danube on horseback, but who probably dismounted to attack the Balkan farming settlements because their horses were too skittish to be ridden into a raid. And it carried on with the Yamnaya, who bred horses genetically very close to the domesticated kind that replaced most wild ones.

The Yamnaya's horses might have looked like the modern Przewalski's, Anthony thinks: stocky, with short legs, big heads and thick necks. They would probably still have been hard to control, compared to the feathered steeds that would pull Hittite war chariots fifteen hundred years later, and certainly compared to the horse that a Neapolitan nobleman would put through the paces of dressage in the sixteenth century CE, 'with good art and true discipline, and with a pleasant bridle'.[6] But they were likely compliant enough for the purposes of herding cattle and sheep. And that, for Anthony, is what counts.

Because the Yamnaya corralled their animals on horseback, much like the Kalmyks in the nineteenth century, he believes that they could manage far larger herds than their ancestors had been able to. This meant that they had animals surplus to their basic requirements, and the surplus animals could be exchanged in tributes. Through tributes the Yamnaya built alliances and extended their influence. Their scouts, being mounted, could also roam further in search of new pastures. Having mastered the miracle of speed, Anthony claims, the original *perekati-pole* went on to dominate large swathes of Eurasia – and their language did the same.

There was once a culture that we call Yamnaya. There was once a language that we call Proto-Indo-European. Why do experts feel so confident that the people of the Yamnaya culture spoke Proto-Indo-European? They can't claim this with any certainty, since the

Yamnaya didn't write their language down, but they can come close. A language mirrors its speakers' world. From the broken reflection of that world that linguists have pieced together, archaeologists and geneticists can tell which prehistoric people were most likely to have inhabited it.

Proto-Indo-European has been reconstructed based on a comparison of its offspring, an exercise that has revealed the Indo-European sound laws – those shifts in pronunciation that transformed the mother tongue into its daughters. Generations of historical linguists have laboured on this task. No other proto-language has received as much scholarly attention, which is why we know more about it than any of its contemporaries.

We know, for instance, that Proto-Indo-European was a highly inflected language, meaning that the role a word played in a sentence was signalled by modifying the word itself. In an analytic language, by contrast, meaning is conveyed by the order of the words, or by recruiting additional parts of speech such as prepositions ('at', 'on' and 'to' are English examples). In fact English is a particularly poor indicator of what its ancestor was like, because English has been transforming itself over hundreds of years from an inflected language to an analytic one. Older Indo-European languages, such as Sanskrit, Ancient Greek and Latin, retain the inflected character of Proto-Indo-European, and a comparison of Latin (inflected) and English (analytic) sentences reveals the difference between the two types. *Canis mordet puellam* means 'the dog bites the girl' in Latin. The *-is* ending of *canis* and the *-am* ending of *puellam* indicate that *canis* is the subject of the verb and *puellam* the object. In the modified sentence *canem mordet puella*, the dog and the girl haven't switched places but they have switched roles, giving 'the girl bites the dog' – a reversal of fortune that can only be expressed in English by shuffling the words around.

The reconstructed Proto-Indo-European lexicon consists of about sixteen hundred words. Strictly speaking, most are stems rather than words, where a stem may or may not be a word in its own right, but it hatches a brood of them once it has been massaged by the rules of grammar.⁷ This vocabulary represents a mere skeleton of the language as it was once spoken, but it's still an impressive feat of resuscitation given that the language in question has been dead for more than four thousand years. Armed with it, we can sketch a word-portrait of its speakers.

First, though, a point of order. Until now, for simplicity's sake, I have stripped my transcriptions of almost all phonetic symbols and diacritics – those accents, umlauts and other marks that indicate how a letter should be pronounced. This won't do going forward, because Proto-Indo-European used a different palette of sounds from English, and those odd markings are needed to capture them all. From now on, then, they will be restored to their rightful places. My advice is not to fixate on them, because the point should be clear even if a word's pronunciation isn't, but if you'd like to know how Proto-Indo-European sounded, the 'Note on pronunciation' at the end of this chapter explains the code in full.

Some Proto-Indo-European words sound utterly foreign to speakers of living Indo-European languages, even though they can be transformed into their modern counterparts by the application of the appropriate sound laws. Others look and sound familiar, but those ones are almost more disconcerting because they give the false impression that you might be able to hold a meaningful conversation with a Proto-Indo-European, were you to meet one in the street.⁸ Take *dhugh₂ter-, which has been reconstructed from English 'daughter' and its counterparts *duhitár-* in Sanskrit, *thugátēr* in Greek, *dustr* in Armenian and *duktė* in Lithuanian (note the asterisk signalling a reconstruction, a hypothetical word that itself has never been documented). It joins *ph₂ter- (father), *meh₂ter- (mother), *bhreh₂ter- (brother), *swesor- (sister) and *suHnu- (son) to make up the nuclear family.

These terms belong to the stable core of the lexicon, those words that have changed relatively little over time and space, perhaps because humans are so intimately concerned with the things they represent that they find misunderstandings regarding them intolerable. Besides kin, they cover certain animals, the parts and functions of the body, elemental concepts such as water, wind and night, and numbers. (Linguists aren't sure that Proto-Indo-European had a word for 'one', but the next four number-words have been reconstructed as *duó-, *tréy-, *kʷétwor-, *pénkʷe.⁹) Yet even these universally important words hint at who the first Proto-Indo-European-speakers were — at what mattered to them.

Variations on *h₁ek'wos or 'horse' (Sanskrit áśva-, Greek híppos, Latin equus) and *k'won- or 'dog' (Sanskrit śván-, Greek kuón-, Latin canis) crop up reliably from Galway to Calcutta.† *k'erd-, the root of both 'cardio' and 'heart', sometimes occurred in the context of an expression, *k'red dheh₁-. Literally meaning 'to put your heart', this became śrad dhā- in Sanskrit (believe) and crēdō in Latin (I believe). A word meaning 'star', *h₂ster-, shines steadily through all its descendants. Waypoint for night travellers since all humans were African, it was known to Sogdian merchants on their camels as stārē, to homebound Odysseus as astḗr, and to Icelanders fishing for herring after dark as stjarna.

Taboo words, on the other hand, are recognisable by their very *in*stability. Because they were unsayable, because they were forever being circumlocuted and euphemised, they saw a high turnover (but left traces of themselves in obsolete phrases and personal names, like snakes shedding their skins). By way of analogy, think of how a tired four-letter word in English, beginning with an *f* and referring to a sexual act, became 'frick', 'freak' and 'fudge'.

† Note that in Sanskrit the sound represented by *ś* resembles English *sh*. In the alphabets of the living Baltic languages — as well as in long-dead Hittite — *š* represents the same sound (this will become relevant in later chapters, where I will occasionally replace both *ś* and *š* with *sh* to highlight similarities across languages).

Early Indo-Europeans might have been surprised by some of our hang-ups, but they had hang-ups of their own. Certain dangerous animals were taboo for them, including bears, wolves and that perennial symbol of evil, the snake. The brown bear was still being euphemised in historical times. Slavic-speakers called it *medŭvědĭ* (honey-eater), while for Dutch-speakers it was *bruin* or 'brown'.

Proto-Indo-European had a rich vocabulary of dairying, including 'cow' (*$g^w ow$-), 'sheep' (*$h_3 owi$-) and 'to milk' (*$h_2 melg'$-), as well as words meaning 'sour milk', 'buttermilk' and 'cheese'. It had a word for 'foal' (*$polH$-), possibly one for 'tame' and another for 'wool'. There were words for 'honey' (*mélit, the root of French *miel*, English 'mellifluous' and the Ukrainian city of Melitopol), 'hornet' and 'wasp', though curiously not for 'bee'. The language contained a smattering of terms relating to crops and land cultivation, including one for 'plough' (*$h_2 erh_3$-, the root of the Latin verb *arare*, 'to plough', and of English 'arable'). But these amount to a fraction of the agricultural vocabulary that has been reconstructed for Proto-Afro-Asiatic, a language known to have been spoken by farmers.

Arguably the most controversial word in the Proto-Indo-European lexicon is *$k^w ék^w los$. This was reconstructed from words meaning 'wheel' in the daughter languages, including Sanskrit *cakrám*, Greek *kúklos* and Old English *hwēol*.† Some argue that *$k^w ék^w los$ could initially have referred to something abstract: a circle or round. Human beings have always known things that turn, after all, starting with the sun, the moon and the menstrual cycle, and it might only have acquired its association with a means of transport later on (just as 'mouse' acquired a second meaning of a device

---

† To say *hwēol*, work up a head of air before that *h*, take a running huff at it. The *hw* sound is rarely heard in English any more (you see its silent tombstone in the spelling of *wheel*, which when spoken sounds exactly like *weal* or *we'll*), but both letters were once pronounced.

for steering your computer cursor). However, because a second 'wheel' word has been reconstructed (*$roteh_2$*, as in 'rotate'), along with others for 'axle' (*$h_2ek's$*) and 'to convey in a vehicle' (*$wegʰ$*), most linguists are satisfied that it already designated the round wooden object, first solid, later spoked, that made a wagon move. In other words, they believe that the speakers of Proto-Indo-European knew wheeled transport.

Already, then, the lexicon is constraining, geographically and temporally, who those speakers could have been – helping us, at least, to say who they could *not* have been. People who spoke of wheels and wagons could not have lived before 3500 BCE, when that technology was invented. It was around the same date that dairying became common on the Eurasian steppe. But in 3500 BCE, according to the archaeological record, the only steppe-dwellers who herded horses, cattle and sheep lived to the west of the Urals (the Botai, to the east, had horses but no cows or sheep). By the time other people emerged who kept that combination of animals, a thousand years later, linguists say that Proto-Indo-European was extinct. It had been replaced by its daughter languages.

The Yamnaya are looking like a good fit for the speakers of Proto-Indo-European, then. East of the River Don, where Shishlina excavates, they grew no crops. (Their teeth confirm this, since they are strikingly free of caries – a sign that starchy cereals were not a major component of their diet.) But the oldest known Yamnaya remains come from further west, closer to the Dnieper. In that region, which falls within the *chernozem* belt, they may have stayed still for long enough to plant and tend edible plants, which then supplemented their diet of meat and milk. And they may have used primitive ploughs called ards to do so ('ard' also derives from the Proto-Indo-European word for 'plough', *h_2erh_3*-). Even before the Yamnaya, steppe herders living close to the farmers of eastern Europe took up their techniques in a small way, so it wouldn't be surprising if some Yamnaya had too, in which case they would have needed words to describe what they were doing. The presence of a word for 'wool' in the reconstructed lexicon need not pose a problem, since it could have referred to the sheep's short underwool rather than the product of long-haired animals that only reached the steppe much later. Even the lack of a word for 'bee' can be explained, if the Yamnaya, inhabitants of those treeless expanses, traded honey in from their neighbours in the forest belt. They would have known the product, in that scenario, but not necessarily the insect that made it.

There is one puzzling lacuna. The Yamnaya knew copper and bronze, but no metal word has been reconstructed for Proto-Indo-European. An absence is less troubling than the presence of a word that doesn't fit the archaeological profile, however, because the reconstructed lexicon is far from complete. In all languages words are replaced over time, and if there is no written record of older forms then it might not be possible to reconstruct the ancestral one. So Proto-Indo-European might have had a word for copper that linguists have simply been unable to reconstruct. For the same reason, it's not always possible to say whether a given word goes all the way back to the proto-language, or only emerged in a daughter branch (since it might not be 'visible' in all the branches that had it). Historical linguists are aware that their reconstructed lexicon is a palimpsest. That is, it combines the vocabulary of several early Indo-European languages spoken over a span of time by a series of different but related cultures. And a rather threadbare palimpsest at that.

The lexicon is nevertheless revealing about how early Indo-European societies organised themselves. *wedh meant both 'to marry' and 'to lead (a bride) away', while *potis meant both 'husband' and 'master' (from *dems-potis, 'head of the household', comes English 'despot'). There is a panoply of words for a wife's in-laws, but none for a husband's. Linguists infer that a woman moved into her husband's household upon marriage, and that wealth passed down the male line.

pek'u, the root of 'pecuniary' but also (via Grimm's Law) of 'fee' and 'feudal', referred both narrowly to livestock and more broadly to wealth. Wealth could be inherited and received in tribute, or in exchange for a bride, but it could also be stolen. There are plenty of words relating to raiding and booty, and plenty on the themes of restitution and blood price. The person who adjudicated disputes was the *h₃rēg's*. The root of Latin *rēx*, Sanskrit *rāj-* and Gaulish *rīx*, this came to mean 'king', but to begin with it might have referred to someone more like a priest: one who stood for what was 'right' or who 'regulated'.

A hierarchy existed in which wealthier, more powerful clan chiefs received tributes from their subordinates in return for protection and favours. All of this was coded: there was a word for an 'oath' (*h₁oitos*), one for a 'follower' and another for a 'service'. And that relationship, between a mortal patron and his mortal client, was mirrored in the relationship between humans and gods. Speakers of Proto-Indo-European paid tribute, in the form of sacrifices, to a pantheon of deities presided over by Father Sky, *dyēus ph₂tēr*.

The vertical warp of the hierarchy was reinforced by a horizontal weft of alliances, and both were sustained through hospitality: *ghostis*, or 'guest-friendship'. It helps, in this case, to say the word out loud: the *gh* signals an aspirated sound, somewhere between the *g* of 'guest' and the *h* of 'host'. *ghostis* combined both concepts. A person who had enjoyed another's hospitality was duty-bound to return it, and guest rights were extended to strangers. Linguists deduce this because *ghostis* is the reconstructed root of Gothic *gasts* (guest) and Old English *giest* (stranger, guest), but also of two Latin words: *hospes* (host, guest) and *hostis* (stranger, enemy). *Hospes* underlies the word 'hospitality' while *hostis* is the root of 'hostility', via Latin loans to English. These two words seem antithetical in their meaning, but they were once linked by the concept of a stranger, that passer-by whom a good welcome might turn into a friend and a bad one into a foe.[10]

ghostis might have been the unspoken rule that allowed speakers of Proto-Indo-European to pass safely through each other's territory. It would have become more important over time, as what started out as a small, tightly knit community grew and dispersed. Some claim that *ghostis* contains the echo of another word, *ghes*, meaning 'to eat'. If they are right, then from the beginning the proper way to receive a guest may have been to lay on a feast. At the feast, where they drank beverages made from fermented milk, plants or honey (*medhu*, 'mead'), the guests listened as a bard praised the generous host (*ghosti-potis*, literally the 'lord of guests'). Bards held a hallowed status in Indo-European society. They could make or break reputations, build a person up with blandishments or slam him down with satire. And for Indo-Europeans, reputation was everything. A warrior sought fame on the battlefield, because with fame came immortality. The same phrase, 'fame imperishable', appears in *The Iliad* (*kléos áphthiton*) and the *Rig Veda* (*śrávaḥ ákṣitam*), suggesting that the concept was familiar to the speakers of Proto-Indo-European too.

The bard sang of cattle, death, theft and glory, and of heroes, gods and kings. He sang from memory, drawing on stock phrases whose meaning his audience understood. A hero was one who urinated standing up; a horse was always swift; 'bipeds and quadrupeds' was shorthand for all the beasts of the Earth. Not only did these formulae serve as mnemonic devices, they may also have had the power of charms. Linguists look for them when they're trying to establish if two myths are related, if they grew out of the same poetic tradition.[11] They are part of their toolbox for reconstructing the myths that the first Indo-European poets told, as the lord and his guests sipped mead or smoked cannabis beneath the stars. Those myths can be as informative about who they were as their words for barley or bronze.

One such myth recounts how the world came to be. Almost every human culture has a creation myth and the speakers of Proto-Indo-European were no different. Often such a myth involves the world being fashioned from a primordial creature, but in their version, uniquely, a primordial Man (*Manu*) conjures the world from his Twin (*Yemo*). After journeying through the cosmos in the company of a primordial cow, Man and Twin decide to make the world. To do this Man must sacrifice Twin and cow so that he can build it from their dismembered parts. The sky gods and goddesses help him, and once the world exists Man becomes the first priest, overseeing ritual sacrifice. The gods then create Third Man (*Trito*), to whom they give cattle, but a monster serpent steals the cattle. The name of the serpent is *Ngʷhi*, which is a literal negation: 'not'. Fortified by an intoxicating beverage and assisted by a god, Trito overcomes the serpent and frees the sequestered cattle. He then makes a gift of cows to Man the priest, so that the sacrifices may continue and cosmic order be restored. Trito is the first warrior, and that first cattle raid is his initiation into manhood.

The idea of a serpent keeping something vital from humans, usually water, is not exclusively Indo-European. You find it on every land mass, even in cold places where there are no snakes (indigenous inhabitants of snake-free Alaska tell of a serpentine sea monster called a *palraiyuk*). Some mythologists locate the trope's origins in Africa, from where they say it spread to all the other continents with the first long-distance migrants. Even after an ice age had caused people to retreat, and subsequent diasporas had scattered them again, the memory of a dangerous reptile stayed with them.

Nor is the myth of a hero destroying the serpent exclusively Indo-European, although Indo-European versions of it are probably the best known (similar myths were told in ancient China, Egypt and sub-Saharan Africa). In Trito lies the germ of every Indo-European dragon-slayer, from Siegfried who slew the dragon on the glittering heath, to Bellerophon triumphing over the Chimaera in *The Iliad*, Fergus mac Léti, King of Ulster, decapitating the *muirdris* in Loch Rudraige, and the great dragon-slayer of the *Rig Veda*, Indra. 'Let me now sing the heroic deeds of Indra, the first that the thunderbolt-wielder performed. He killed the dragon and pierced an opening for the waters; he split open the bellies of mountains.' Often there is a chink in the hero's invulnerability: Achilles' heel, a spot on Siegfried's back the shape of a linden leaf, the Persian hero Esfandiar's eye. The chink is the opening through which death finds him. It is the reason he never outlasts the dragon for long.

Some scholars wonder if the sacrifices that shamans performed at Khvalynsk, in the fifth millennium BCE, or at contemporary settlements overlooking the Dnieper Rapids, were re-enactments of the dismemberment of the primordial cow. They think it possible that the Yamnaya inherited those rituals along with the myth that inspired them. And there are two other themes in the Indo-European creation myth that are potentially reflected in the archaeological record: a moral justification for raiding, and brotherhood. Archaeologist James Mallory has suggested that the myth of Trito was the psychic motor that drove early Indo-European expansion, because it legitimised raiding. Cattle were given to *us*, it implied, because only we know how to sacrifice them properly, so if anyone else has cattle, they stole them. Raiding was the perpetual response to the original insult.

Brothers are the protagonists of the Proto-Indo-European creation myth, and bands of brothers are central to all later Indo-European mythologies. The Germani had the *Männerbund* and the Irish the *fían*, while the Romans had the *luperci*. The brotherhood took as its symbol the migratory, predatory wolf (the symbol is preserved in the name *luperci*, from Latin *lupus*, 'wolf'), and it was associated with nakedness, promiscuity and war. Upon leaving the home of their parents these young men died in the eyes of society and became outlaws. Before engaging in combat, they entered a sort of battle frenzy (*berserksgangr* in Norse, *ríastrad* in Irish), sometimes triggered by imbibing an intoxicating substance. After seven years of hunting and raiding in the wilderness, they returned to take their places in society as married men and warriors. They were reborn.

Underpinning the brotherhood concept was that of the age-set, all those sons born to the elite in a given year. The age-set is well documented in historical Indo-European-speaking societies – among the Gauls, for example, as well as the Romans and Greeks. It was a way of building *esprit de corps* within a birth cohort, since it imposed a rhythm on men's lives: boys grew up together, were initiated as warriors together, married and became grey-haired sages together. It also regulated population growth and pressures on inheritance by fixing the age of marriage (about twenty-one) and hence the length of a generation. In such societies, it was these same-age sons, these symbolic twins, who as teenage would-be warriors were responsible for most of the raiding. They pushed out the territorial frontiers.

The age-set concept might even predate history, if the oldest known stories told by Indo-Europeans reflect even older, prehistoric customs that they inherited. One such story is the Hittite founding myth – a myth, that is, that speaks to how the Hittites understood their own origins. At the time that their records began, about 2000 BCE, they controlled the city of Kanesh (modern Kültepe) in central Anatolia. The myth begins: 'The Queen of Kanesh bore thirty sons in a single year.' One way of interpreting the queen's unfeasibly large brood would be to think of it, not as a biological brotherhood, but as a social one: an age-set.

There is evidence that the age-set, and the warband, were more than mythical inventions. In the Russian steppe, there are clusters of kurgans where young men of about the same age (as revealed by their bones) but of different families (as revealed by their genes) were buried together. They may have been warbands cut down in their prime. As we'll see later, archaeologists have even found probable remains of initiation rites that involved the sacrifice of dogs and wolves. These foreshadow the Roman festival of Lupercalia by more than a thousand years.

The age-sets were long assumed to have been all-male. They tended to be, in myth, but there were exceptions to the mythological rule, and archaeologists have stumbled across what might have been real-life exceptions too. In the Bronze Age cemeteries where the putative warbands were buried, it has happened that a skeleton initially identified as male, on the basis of bone measurements, has turned out to be genetically female. Like the young men buried close to them, these young women were strikingly tall and athletic. Their presence has, unsurprisingly, inspired no end of speculation. The warbands had a reputation for sexual voraciousness, so could they have been priestesses who oversaw fertility rituals? Talismans brought along for luck? Or were they initiates themselves, perhaps the fearless only daughters of clan chiefs?

Much later, the Iranic-speaking Scythians who dominated the Pontic steppe in the last millennium BCE left behind kurgans that are indisputably the graves of female warriors. A handful of ancient historians, including Strabo and Plutarch, describe a caste of female fighters called Amazons who lived among the Scythians, while the fierce maidens known as valkyries are a staple of Norse mythology. We don't know how much of these stories is fiction and how much is fact, but the Vikings, too, buried female warriors with horses and swords. It's possible that the Bronze Age nomads of the steppe had a more flexible concept of gender than they have generally been credited with, and that the girl-women buried alongside their brothers-in-arms, more than four thousand years ago, were the first in a line of female fighters who inspired the Amazons and valkyries of legend.

Sometimes, when a community was under strain, perhaps due to overpopulation, disease or drought, there was no place for the young renegades to come back to. Roman and Greek historians describe what happened under these circumstances, in later periods. A whole age-set might be sent out to make a new life for itself, to find new lands and new wives (or husbands). The notion of banished youth is there in the Maruts, the violent siblings, 'eager for fame', who accompany the warrior-god Indra in the *Rig Veda*. It's there in Romulus, who, having murdered his twin and founded Rome, gave the signal to his unmarried followers to kidnap the Sabine women. And it's there again in the Hittite founding myth. Having given birth to her thirty sons, the Queen of Kanesh sent them off to a foreign land. She went on to have thirty daughters, whom she raised herself at Kanesh. Later on, the grown-up sons and daughters met each other on the road and, not recognising each other, slept together.

The early Romans even had a term for the age-set that was cast out: *ver sacrum*, or sacred spring. But the concept was known, with variations, across ancient Europe and perhaps further afield. 'When the lands of their birth could no longer take them, the Gauls in their abundant multitude sent three hundred thousand men like a sacred spring to seek new abodes.'[12] If the speakers of Proto-Indo-European were familiar with the concept, one could speculate that those who struck out from a river valley on the Pontic steppe, carrying the Y chromosomes of a single clan or tribe, might have been a sacred spring banished by the *$h_3rēǵs$*. Embracing the nomadic life, they did what was expected of them and found a new home. So successfully did they adapt to that home that millions of people alive today carry their genes and speak versions of their tongue. We *can* only speculate about the events that forged the Yamnaya, for now, but as to what language they spoke, scholars have reached a broad consensus: it was Proto-Indo-European. Every time we speak or write in a descendant of that language, we unwittingly scatter clues as to who those nomads were and what they believed. They live on through all of us who speak Indo-European today.

Note on pronunciation

The languages mentioned in this book were written in a variety of scripts or alphabets, if they were written at all. Those that were written in scripts other than the Roman one, or its variants, are transliterated here into the latter. Proto-Indo-European was never written down, but linguists have devised a system of recording the sounds that they think it used, by adding diacritics and some extra letters to the Roman alphabet. This Proto-Indo-European alphabet has often been revised, and parts of it remain controversial, but what follows is an abridged description of it, to help readers pronounce the reconstructed Proto-Indo-European words mentioned in this book (I won't explain the alphabets of any other languages, except where doing so is key to an argument).[13]

Proto-Indo-European is thought to have used the vowels *a*, *e*, *i*, *o* and *u* in their long (*ā*) and short (*ă*) forms. It used the semi-vowels *y* and *w*; the nasals *m* and *n*; the liquids *l* and *r*; the sibilant *s*; and a rich assortment of other consonants that can be arranged in the following triads: labials (using the lips) *p b bh*; dentals (using the teeth) *t d dh*; two kinds of dorsals (in which the back of the tongue touches different parts of the roof of the mouth) *k g gh* and *k' g' gh'*; labio-velars k^w (a 'rounded' k, a bit like English *qu*) $g^w g^w h$. The *h* after another consonant indicates that it is aspirated, meaning that you should expel air as you make the first sound. In addition to these consonants there were also the laryngeals, a type of consonant that was only preserved in the Anatolian branch of the family. There are generally thought to have been three of these

(h_1, h_2, h_3), because they 'coloured' the sounds around them in three different ways, but their pronunciation is debated. h_1 might have sounded like the *h* in English 'host' and h_2 like French *r* or Arabic *q*, with the back of the tongue near the uvula (the flap of soft tissue that hangs down at the back of your throat). h_3 has been described as a voiced version of h_2, meaning that it required the vocal cords to vibrate. A capital *H* indicates that the variety of laryngeal is not known. Somewhere between a vowel and a laryngeal was schwa (ə), which is also very common in English (it is the terminal sound in 'the', for example). An illustration of schwa's in-betweenness is the Proto-Indo-European word for 'father', which has been written in two different ways: *$ph_2tēr$ and *pətēr. Finally, accents show where the stress falls. In Proto-Indo-European, the stressed syllable is thought to have been of higher pitch than the others (as it was in Ancient Greek), rather than louder (as it is in English).

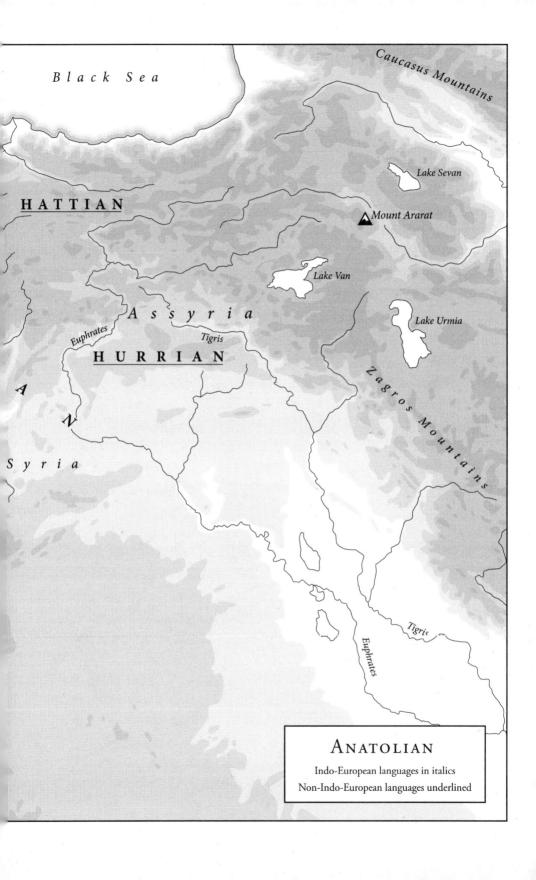

3

First Among Equals

Anatolian

nu ninda an ezzatteni wadar – ma ekutteni

Bedřich Hrozný stared at the passage. Next to the wedge-shaped marks that represented the word *ninda* was an ideogram – a character representing an idea – that he recognised from Sumerian texts. There it meant 'bread', so perhaps it did here too. The repeated suffix *–(tt)eni* suggested a verb ending, one tantalisingly close to verb endings in Latin and Sanskrit. This gave him the husk of the phrase: 'bread x-ing, y z-ing'. But Hrozný knew that he could do better than that. In Old High German *ezzan* meant 'to eat' (the word became *essen* in Modern German). If such echoes were meaningful, then following the theme hinted at by the bread ideogram, *wadar* might mean 'water'. Putting it all together he came up with: 'now you will eat bread, further you will drink water'.

Hrozný had just broken the code of the first Indo-European language ever to be written down: Hittite. Though he didn't know it at the time, the document that had given him the key was the last will and testament of an early Hittite king, Hattushili I, who had ruled in north-central Anatolia in the seventeenth century

BCE. Hattushili was a master of spin, especially when it came to himself: 'his frame is new, his breast is new, his penis is new, his head is of tin, his teeth are those of a lion, his eyes are those of an eagle, and he sees like an eagle'. But he was also an accomplished warrior who had laid the foundations of one of the great empires of the preclassical world. As he lay dying he dictated his plan for his succession, but in the ancient equivalent of the microphone being left on after the interview has concluded, an over-enthusiastic scribe kept scribbling and captured his last words.[1] As death rushed up to meet him, Hattushili the Lion was seized by terror: 'Wash my corpse well! Hold me to your bosom! Keep me from the earth!' Three thousand years after its ancestor was first spoken on the shores of the Black Sea, the first Indo-European cry to reach us is heartrending in its humanity.

The story of Hrozný's decipherment is only a little less so. He cracked the code in 1915 and quickly announced his preliminary conclusion that Hittite was an Indo-European language. The Czech linguist was in a hurry. He had just become a father. The First World War was sixteen months old and, though he was extremely short-sighted, he knew that he could be called up at any moment. He wanted to secure his academic reputation. There was also competition to decipher Hittite, ever since rumours had begun to circulate, more than a decade earlier, that despite being written in the same cuneiform script as the Akkadian language spoken further to the south and east, it belonged to a different family.[2]

So far the rumours had not been taken seriously. At the turn of the twentieth century, relatively little was known about the Hittite Empire. It was only in 1906, thanks to excavations near Bogazköy, about a hundred and fifty kilometres (ninety miles) east of Ankara, that its impressive capital, Hattusha, had come to light. Conventional opinion held that none of the great Bronze Age civilisations, not Egypt nor Babylonia nor Assyria nor the Hittites, had spoken Indo-European. The Phrygians were thought to have been the first speakers of Indo-European on the Turkish peninsula, starting from the twelfth century BCE. (The rumours had nevertheless filtered out beyond academia. Austrian writer Robert Musil threaded them into his famous novel *The Man Without Qualities*, which is set in Vienna on the eve of the First World War: '. . . in his eyes even the ladies and gentlemen of the highest society performed a significant if not readily definable office when they chatted with learned experts on the Bogazköy inscriptions . . .'.[3])

The 1906 expedition to Bogazköy (now Boğazkale) had turned up thousands of inscribed clay tablets which the archaeologists realised were the Hittite royal archives. It was these documents, which appeared to have survived a great blaze, that had allowed Hrozný to take his first steps towards deciphering the language and demonstrating its Indo-European credentials. Before the war, he had been sent to Constantinople (Istanbul) to remove them from their packing cases, clean, sort, photograph and copy them. Having returned home to Vienna, he announced his preliminary conclusion to a chilly reception from his peers. He realised that he would have to supply more evidence. Then he was called up.

Luckily for Hrozný, as a bespectacled and somewhat bookish recruit to the Austro-Hungarian army, he was sent back to Constantinople. There he came under the relaxed command of one Lieutenant Kammergruber, who allowed him to spend his days at the Imperial Ottoman Museum, insulated from the chaos of the wartime city by heaps of fire-hardened clay. Hrozný took care to thank Kammergruber when he published his more thorough findings in 1917. This time his fellow Indo-Europeanists took him seriously, but they also expressed their discombobulation before the sheer strangeness of Hittite, a language that is more different from English than Urdu or Albanian, while still, as they would eventually acknowledge, belonging to the same family.

Hittite isn't alone in the Anatolian branch of the family, though it is the best-documented member of that branch. Other long-dead Anatolian languages would be identified later, including Palaic (once spoken in northern Anatolia), Luwian (thought by some to have been spoken at Troy) and Lydian (thought by others to have been spoken at Troy, but also by 'rich as' Croesus, the last king of Lydia). A new Anatolian language was discovered at Hattusha as recently as 2023, on a clay tablet unearthed during that summer's excavation, but it has yet to be deciphered.[4] The main reasons for classifying the Anatolian languages as Indo-European are the ways that they inflect nouns and verbs to signal their role in a sentence (including those verb endings that helped Hrozný to decipher Hattushili's will). But there are also many ways in which the Anatolian branch differs from the other Indo-European languages. It is the only one to preserve de Saussure's lost consonants, the laryngeals. Hittite verbs have two tenses, past and present; other ancient Indo-European languages have up to six. Hittite marks nouns according to whether they are animate or inanimate only; other Indo-European languages further divide the animate category into masculine and feminine.

From the outset, the Anatolian branch struck linguists as old; as capturing an archaic, even nascent state of Indo-European. They disputed where it stood in relation to Proto-Indo-European, but the prevailing view was that it was the eldest daughter, the first to split from the Proto-Indo-European trunk. All agreed that it had to have been born very close to the birth of the entire family, meaning that any theory of the origins of Indo-European had to explain the origins of Anatolian too, in all its glorious eccentricity. That task would occupy scholars throughout the twentieth century and into the twenty-first.

Through the sitting-room window I'm watching dusk thicken the cascading branches of an aged cedar tree. Over the wall, from next door's garden, looms a giant sequoia. I'm in Cambridge, England, at the home of a giant of twentieth-century archaeology, Colin Renfrew. Lord Renfrew of Kaimsthorn, to give him his proper title, is explaining that the exotic flora were imported by a former master of Jesus College with a passion for botany, when suddenly the door opens and in comes Jane, Lady Renfrew, balancing a teapot on her walking frame. She is also an archaeologist, one who in her day specialised in the use of plants in prehistory. Soon we're sipping tea, eating sponge cake and discussing another titan in their field, Marija Gimbutas.

Even before the twentieth century, archaeologists knew that there had been two major cultural transformations since the last ice age, which had affected both Europe and Asia. The first was the farming revolution. The second was triggered by those nomadic herders who roamed the western Eurasian steppe about five thousand years ago. In neither case could they prove that the cultural shift had been accompanied by migration. The new ways of life could simply have been taken up by existing Europeans and Asians, without the need for any movements of people. But if new cultures could diffuse through populations, so in theory could new languages. There was general agreement that a transformation as dramatic as the spread of the Indo-European languages could not have happened in a vacuum. It had to have piggy-backed on some major social upheaval, and that upheaval must have left a trace in the archaeological record.

Gimbutas was the leading proponent of the theory, discussed in the last chapter, that the steppe nomads had spread those languages. The 'steppe hypothesis' had been aired since the nineteenth century, but she did much to develop it after the Second World War when she herself was rootless. Born in Lithuania, she had been forced to flee by the Soviet occupation of her country, eventually settling with her family in the United States. By the time she took up a post as professor of European archaeology at the University of California, Los Angeles (UCLA) in 1964, her theory was fully formed. In her conception of the past, southeastern Europe had once been home to peaceful, mother-centred farming societies who had practised a goddess cult. These societies she referred to collectively as 'Old Europe', because they had been founded by the first farmers to settle the continent.[5] Starting around 4500 BCE, several waves of horse-breeding nomads had come west from the steppe, entering the farmers' orbit. The invaders were male, aggressive and patriarchal, and in the ensuing culture clash Old Europe ceased to exist.

When Gimbutas presented this theory at UCLA, she found that she was out of synch with her fellow archaeologists. A new paradigm had emerged after the war, according to which diffusion, not migration, was the dominant driver of change in prehistory. This 'new archaeology' rejected the simplistic equation of flux of people with flux of language, the error, as its proponents saw it, of the discredited archaeologists associated with the Nazi regime. Some of the leading thinkers in the new paradigm were at UCLA, but it fell on fertile ground all around a world traumatised by the memory of the Holocaust. Jane recalls the first time she saw Gimbutas, at a conference of prehistorians in Prague in 1966: 'She was very elegant and she had wonderful red hair. She was wearing this beautiful green dress, and she was talking about Indo-Europeans. After a few minutes, once she'd really got going, Professor Hawkes [Christopher Hawkes, then professor of European archaeology at Oxford University] got up and banged on the desk. He said, "Don't take any notice of this woman, she's talking absolute nonsense," and sat down again. She said, "Oh my good friend Professor Hawkes and I disagree, but I'm just going to carry on," and she did.'

Jane sighs happily.

'She took a great deal of flack,' says Colin.

Gimbutas was older than Renfrew by a generation, but they came together through their shared interest in Old Europe. He was an authority on radiocarbon dating, she was a polyglot with deep knowledge of the languages and folklore of eastern Europe. They respected each other's knowledge and energy and collaborated on several projects. In 1967 she invited him to spend a term at UCLA, lending him and Jane a house that she owned in Topanga Canyon. While he was there he came under the influence of the new archaeology, and the two of them had lively discussions. In 1987, by which time he had been named professor of archaeology at Cambridge University, he published his own theory of the Indo-European languages: they had originated in Anatolia nine thousand years ago, spreading east and west with farming. The two leading theories of the Indo-European homeland, and the only two to survive into the twenty-first century, were therefore the brainchildren of two friends who had excavated side by side at the same sites.

Archaeologists were more enthused by Renfrew's model. They weren't convinced that marauding bands of steppe nomads could have succeeded in imposing their languages on what was, by five thousand years ago, a large and dense population of farmers, in Europe as in parts of Asia. And they were sceptical of Gimbutas' ideas about the matriarchal organisation of Old Europe. These remain controversial today. Many archaeologists acknowledge that women played different roles in farming and steppe societies. Women were likely more central to food production in the settled societies, for example. Some archaeologists even think that the hundreds of fat-bottomed female figurines that have been extracted from Old European cemeteries reflect a reverence for the generations of women who coaxed nature into providing new foods. But even if farming societies worshipped female deities, archaeologists then and now felt that those societies were as male-dominated and warlike as any other in prehistory.

Linguists, on the other hand, never really bought Renfrew's hypothesis. For them the dates didn't work. The oldest texts in the Hattusha archives date to 1650 BCE, but Hittite names and words crop up in documents that are three hundred years older. There are texts written in the other Anatolian languages that are almost as old, and having compared these with the Hittite texts the linguists concluded that by 2000 BCE Hittite and its sisters were already distinct languages. Extrapolating backwards, allowing time for that divergence, they estimated that the parent language, Proto-Anatolian, was spoken around 3000 BCE. If Renfrew was right, and Indo-European-speaking farmers had moved out from Anatolia around 7000 BCE, the language they left behind, Anatolian, would have had to remain more or less intact for four thousand years before it split. Such stability is unheard of in the history of language. Between five hundred and a thousand years is the rule of thumb for the time it takes a language to evolve into a new one, and it is generally agreed that the interval would if anything have been shorter in preliterate, pre-state societies, where there was no writing or official state language to hold back change.

The linguists had plenty of other objections to Renfrew's Anatolian hypothesis. For one, they considered the evidence overwhelming that Indo-European languages were intrusive to Anatolia – that they had come into it from outside. The Hittites' own archives revealed that they were surrounded on all sides by speakers of non-Indo-European languages, including the people thought to have been the indigenous inhabitants of the region: the Hattians. As early as 1920, an American linguist, Carl Darling Buck, noted that although Hittite followed Indo-European rules of word formation, it was awash in 'alien' vocabulary. Much of that vocabulary turned out to be Hattian.

The linguists were much better-disposed towards the steppe hypothesis, according to which several waves of nomadic herders had departed the steppe starting in the late fifth millennium BCE, seeding the Indo-European languages throughout Eurasia. In this model, the first to leave had travelled down through the Balkans, crossing the Bosporus and implanting Proto-Anatolian in Anatolia (David Anthony equated these migrants with the mace-wielding chieftains who had toppled Varna). It made sense to the linguists to think of Indo-European-speakers entering from the west because ancient records attested that Anatolian languages were concentrated there, while being noticeably absent from the east of the peninsula. The linguists agreed with Renfrew that the farmers' languages had probably spread with them, several millennia earlier; they just didn't think that those languages were Indo-European. In their view, the farmers' languages were now lost to us, except for relics such as Basque, having been displaced by the Indo-European languages that came later.

By the time Gimbutas died in 1994 she was famous beyond academia and her books had a large, mainly female following. Her writing on the goddess cult of Old Europe had given a new generation of feminists hope that societies could be organised differently, because they had been once before. The male-dominated academy took a dimmer view. They accused her of abandoning scientific rigour, of letting her imagination run away with her. As far as the Indo-European languages were concerned, however, both her and Renfrew's theories remained on the table. Others had taken up and refined the steppe hypothesis, notably her former student, James Mallory, and Anthony, but the debate had reached an impasse. It took progress in another field to push it on.

Starting in the 2000s, genetic studies (based on modern genes, not yet on ancient DNA) furnished evidence that migration had, after all, been a powerful force in prehistory. It was then that the long-running debate over the farmers was finally cleared up: not just their technology, but they too had moved, and in force. Their ancestry had replaced at least forty per cent of the European gene pool, and in some places close to a hundred per cent. The first hints emerged that they weren't the last major migration to the region either, because modern Europeans carry the genetic legacy of a third population, besides the farmers and the hunter-gatherers they had displaced.

In 2015, two papers were published in the leading journal *Nature* that confirmed those suspicions about a third major influx to Europe. Using two different methods to analyse ancient DNA, they came to the same conclusion: migrants had radiated east and west from the steppe around five thousand years ago, and in Europe their ancestry had replaced up to ninety per cent or more of the gene pool. The mainstream press went to town on the discovery, and with justification, because the last piece of the tripartite puzzle that defines modern Europeans genetically had fallen into place. They remain overwhelmingly, to this day, part hunter-gatherer, part farmer and part steppe nomad. It's hard to overestimate the seismic impact of the advent of first the farmers, and then the nomads. No later movement had anything like their genetic, cultural or linguistic legacies: not the massive migrations set in train by the fall of the Western Roman Empire, nor the displacements that followed the Black Death, the 1918 flu or either of the world wars. Most European men alive today, and millions of their counterparts in Central and South Asia, carry Y chromosomes that came from the steppe.

Some scholars continued to support the idea that the Indo-European languages were born in Anatolia nine thousand years ago, but they were now in the minority and the man who had proposed it was not among them. Giving the inaugural Marija Gimbutas memorial lecture at the University of Chicago in 2017, the by then eighty-year-old Renfrew retracted. In the main thrust of her argument, he said, if not in every detail, his friend and intellectual jousting partner had been right.

When David Reich was a teenager in the late 1980s, he liked to play a game called *Civilization*. Not the computer game that later took the world by storm, but a quaint old board game that knew some considerable success of its own. It was a game of strategy that, for once, was not all about war. Each player was assigned a population around the Mediterranean Basin at the dawn of agriculture. Trading commodities and calamities, purchasing 'civilisation cards' such as philosophy or literacy or medicine, they attempted to expand their empire at the expense of others'. The board was colour-coded according to the boundaries of actual ancient civilisations (Egypt, Crete, Italy . . .), and the winner was the first to make it all the way through the 'archaeological succession table' to 250 BCE.

The game, Reich explained to those of us packing an auditorium in Leiden, in September 2022, was the inspiration for a trio of papers that he and more than two hundred collaborators had just published in another eminent journal, *Science*. Summarising their findings was a map centred, this time, on the Black Sea. The map's border was colour-coded to represent the four genetically defined populations that had occupied the western end of Eurasia ten thousand years ago: Iranian farmers, Anatolian and Levantine farmers, Balkan hunter-gatherers and Eastern hunter-gatherers (those who came west from the Caspian Sea after the last ice age, settling around the Dnieper Rapids). Pie charts showed how these populations had mixed over the following millennia, and coloured arrows between the pie charts indicated how migration might have driven the spread of the Indo-European languages.

There's something a little Chaplinesque about Reich. Diminutive, supple (he frequently perches cross-legged when listening to others), he presents a watchful mask to the world until, every now and then, some emotion wells up and pushes the mask aside. You sense inner steel, which is surely what's required to coordinate the large, multidisciplinary consortia that his kind of science involves, but also the heavy responsibility that he feels for communicating that science. No one is more alive to the troubled history of Indo-European studies than this grandson of Holocaust survivors, whom the ancient DNA revolution catapulted to scientific stardom. And here in Leiden, at a conference of historical linguists, the forty-eight-year-old Harvard geneticist had big things to say. The linguists had invited him to speak before the publication of the *Science* papers, and now they wanted to understand those papers' implications for their field. Beyond the wood-panelled room people were making the most of an Indian summer, swimming in the Dutch city's enviably clean canals or drinking beer in the sunshine. Inside all eyes were fixed on Reich, and many more people were tuning in from around the world.

To begin with, he explained, the four populations coded in the map were far more genetically distant from each other than the populations inhabiting the Black Sea region today. They were as different from each other, in fact, as modern Europeans and Chinese people. Even the Anatolian and Levantine farmers in the lowlands, and their upland neighbours the Iranian farmers, were clearly distinct in genetic terms. They were likely descended from local populations of hunter-gatherers that physical and cultural boundaries had kept apart. Then around 6000 BCE, if not earlier, Iranian farmers started moving north through the Caucasus, first in search of land, and later lured by metal. They interbred with the people they met to the north of those mountains, and by 4500 BCE individuals were being buried at Khvalynsk, on the Volga, who carried two types of ancestry: Eastern hunter-gatherer and Iranian farmer. These people were the descendants of mixed couples: steppe-dwellers who might have resembled modern Finns, and the more recent Caucasian immigrants who probably looked more like Georgians.

South of the Caucasus, meanwhile, everything was also in flux. At the eastern end of Anatolia, Anatolian and Iranian farmers were mingling at last, sharing genes and technologies. Uruk's hunger for metal and timber, and the traffic it drove through the mountains, sucked that mingled ancestry up on to the steppe. It was already there by 3700 BCE, when the burial mounds of the fur-clad oligarchs began casting long shadows over the North Caucasus piedmont. And by the time the Yamnaya emerged a few centuries later, they carried a predominantly two-way mix of Eastern hunter-gatherer and Iranian farmer ancestry, with a dash of Anatolian and Levantine farmer.[6]

The Yamnaya and their descendants had carried that genetic profile east and west, as the landmark studies of 2015 had shown. Mallory, Anthony and many others thought that they had also carried Proto-Indo-European with them, and Reich agreed. But he now considered it unlikely that either the Yamnaya or their steppe-dwelling ancestors had seeded Anatolian in Anatolia. Though his group had detected Eastern hunter-gatherer ancestry in one or two burials in the Balkans, they had found none on the Turkish peninsula. From the ancient bones and teeth sent to them by archaeologists who had excavated there they had extracted plenty of Anatolian and Iranian farmer DNA, but none of the DNA of those hunter-fishers who had once hovered over the Dnieper Rapids. From 4500 BCE to 1300 BCE – from the Neolithic to the dying days of the Hittite Empire – there was virtually no trace of the people who had contributed almost half of the Yamnaya's genetic complement, south of the Bosporus.

What this meant, Reich explained, was that there was no genetic link between the steppe and Anatolia, to mirror the linguistic link between Proto-Indo-European and Anatolian. It was possible that a very small number of steppe immigrants had managed to impose their language on the indigenous population of Anatolia, without having any impact on the local gene pool, but such cases were exceedingly rare in the history of language (it has been far more common for an incoming elite to abandon their language in favour of the one spoken by the masses, as happened, for instance, when the Huns and Mongols invaded Europe). It was also possible that Reich's team had not yet sampled enough prehistoric Anatolian remains to detect those steppe immigrants. But given that they had drawn a blank in over a hundred and thirty samples, when each of those samples tapped into many generations of related people and steppe ancestry is unmissable in Europe, it seemed increasingly unlikely that they had overlooked a significant wave of immigration.

If migration really was the main driver of language change in prehistory, as most linguists seemed to believe, Reich felt it was time to revive a theory that predated Renfrew and Gimbutas. In 1926, an American linguist, Edgar Sturtevant, had claimed that Anatolian was not the eldest daughter of Proto-Indo-European, but its sister. They were the two daughters of an older parent language. Sturtevant's theory was dismissed at the time, but Reich said that the genetic evidence supported it. The ancestral language might have been spoken in the Caucasus between five and seven thousand years ago, by people of predominantly Iranian farmer ancestry. From the Caucasus, his team had traced migrations west into Anatolia and north to the steppe. The first group of migrants could have carried the seed of Proto-Anatolian with them, the second the seed of Proto-Indo-European. On the steppe the northern faction interbred with the local tribes, acquiring Eastern hunter-gatherer ancestry, and in the mouths of their descendants – the Yamnaya – the imported language became Proto-Indo-European.

Between five and seven thousand years ago, in fact roughly in the middle of that window, Varna was thriving. The ancestral language could have been the mother tongue of some of the Balkan coppersmiths' clients from over the sea. It could have been the lingua obscura we met in Chapter One. In Leiden in 2022, Reich wasn't able to say exactly where the speakers of that ancestor had lived, whether it was to the north of the Greater Caucasus, in the mountainous region shared by modern Russia and Georgia, or in the more southerly highlands that today fall within the borders of Armenia, Azerbaijan and Georgia (overlapping with Türkiye and Iran). But he believed that further sampling would narrow it down.

Barely had he finished speaking than urgent questions came like a flight of darts from the floor. This was hardly surprising, given that he had just readmitted the possibility that the homeland of the entire family, including Anatolian, lay in Iran, where Oriental Jones had placed it in 1786, or slightly to the north and west of it, in the neighbourhood of Mount Ararat – where Noah's ark had beached. Most of the linguists seated in that wood-panelled room considered that the roots of Anatolian lay deep in the steppe, and that it had arrived from the west rather than the east. Here was Reich challenging both tenets. But his radical proposal was about to get some support from an unexpected quarter.

At the Max Planck Institute for Evolutionary Anthropology in Leipzig, a global powerhouse of palaeogenetic research to rival Reich's at Harvard, psychologist Russell Gray was putting the finishing touches to his own *Science* paper. Reich and Gray both think about language in evolutionary terms, but their approaches couldn't be more different. Reich works with human DNA. He tracks prehistoric people, then asks how their migration patterns fit – or don't – the spread of prehistoric languages as proposed by historical linguists. Gray works with linguistic data, tracking languages *as if* they were biological organisms – as if they evolved like a virus. The data he uses are the words which, through careful sifting (guided by the sound laws), linguists have determined to have been inherited by a family of languages. Gray has trained an algorithm to detect degrees of similarity and difference across this inherited vocabulary, and the algorithm organises the languages into a family tree accordingly (actually a set of trees, ranked by probability). By cross-referencing the trees with historical texts where they are available, and archaeological and genetic data where they are not, he is able to estimate the chronological dates of important events in the family's life (births, deaths, splits).

By 2022 Gray had been applying this method for twenty years, and the results had consistently confirmed Renfrew's hypothesis: they pointed to a homeland of all the Indo-European languages in Anatolia around nine thousand years ago. This had earned him harsh, even withering criticism from linguists who found that hypothesis to be incompatible with their understanding of linguistic evolution (and who, in many cases, didn't like the idea of languages being compared to viruses). It turns out that Gray, a New Zealander with sparkling blue eyes and a penchant for colourful shirts, has a sense of humour – he calls himself a 'tree-hugger' – and isn't easily discouraged. He shrugged off the barbs and went back to work, using the same methods but new data.

There have been many attempts to gather the shared, inherited vocabulary of Indo-European languages in one place, but each of these dictionaries tends to be quite small. Since some of them are also quite old they contain errors that linguists have only identified as their methods have improved (words that shouldn't have been included, for instance, because they are actually loans masquerading as inheritances). Realising that inconsistencies and poor-quality data could be distorting the trees, Gray's colleague, linguist Paul Heggarty, set about building a new dictionary – or database – with eighty-odd collaborators. The result was much bigger than its predecessors, containing vocabulary from a hundred and sixty-one Indo-European languages – twice as many as Gray had started out with in the early 2000s. Some languages, such as Nuristani and extinct Gaulish, had never been included in a database before, and the vocabulary was selected according to strict criteria.[7] It took Heggarty's team five years to build, and when it was ready Gray used it to generate a new set of trees. These placed the homeland of the Indo-European languages, including Anatolian, south of the

Caucasus – in the region of the Armenian highlands – around eight thousand years ago. In other words, they shifted it east and later than his previous estimates. Many linguists protested, though their protests were a little more muted than in the past, and there was praise for the new database.

When the dust settled, it became clear that a third hypothesis had entered the ring, to compete with Renfrew's and Gimbutas'. Geneticists and at least some linguists were converging on a narrative according to which the parent of the Indo-European and Anatolian languages, the lingua obscura of the first chapter, was spoken in the Caucasus region between six and eight thousand years ago – between 4000 and 6000 BCE, that is.[8] Then in 2024, the Harvard group claimed to have narrowed the birthplace of that obscure ancestor down again (and shifted it slightly northwards). They placed it in southern Russia at a time when Varna was still thriving – about a thousand years, that is, before the Yamnaya emerged.

The debate over the ultimate homeland was far from closed, but even those who disagreed with Reich acknowledged that, for the first time, he had demonstrated a genetic link between the steppe and Anatolia. The genetic, archaeological and linguistic evidence could be reconciled if Proto-Indo-European and Anatolian were sisters, as Sturtevant had argued a century before him. It wasn't uncommon, in 2024, to hear scholars who had been working on the Indo-European puzzle for decades say how glad they were to be alive at a time when science was finally providing answers.

It's a myth that you might recognise, since it exists in many Indo-European traditions: a fertility god flies into a rage and goes AWOL, plunging the cosmos into disorder until he or she is found and placated. In the Greek version, the goddess Demeter is so distressed at the abduction of her daughter Persephone by Hades, king of the underworld, that she lets winter descend on the world. In the Greek, Norse and Indic versions, horses or chariots fly over formidable obstacles to reach the errant god, but in the Hittite version the thing that flies is a bee. A goddess sends the bee to track down the aggrieved god Telipinu. Finding him asleep in a meadow, the bee stings him to wake him up. First bleary, then incandescent with rage, Telipinu goes on the rampage until another goddess calms him down.

Some mythologists suspect that the Hittites inherited the story of Telipinu from an older people who knew neither chariots nor horses as modes of transport. If Gray and Reich are right, those people might have herded goats and sheep in the Armenian highlands, or in what is now southern Russia, before finding their way to Anatolia and bringing the myth with them. The Yamnaya might have inherited the same myth, but they updated it with horses – since they had tamed them – before carrying it into Europe and greater Asia. Their descendants added chariots once those had been invented.

Thus language and mythology provide clues to the prehistory of peoples, in this case the Hittites. To take another example, Hittite has a word for 'wheel', *ḫūrkis*, that has a different origin from either of its counterparts in Proto-Indo-European, *$k^w ék^w los$ and *$roteh_2$. That makes sense if Proto-Anatolian and Proto-Indo-European split from an older language spoken by people who did not know the wheel and so had no word for it. Each of the two daughter branches acquired their wheel words later, after they had separated. Logically, then, those words bore no relation to each other.

Not all the linguistic evidence supports a Caucasian origin, though. If carriers of Iranian farmer ancestry spoke the ancestral tongue, you'd expect that language to contain a good number of farming-related words, which it would have bequeathed to its daughters. Yet the reconstructed vocabularies of Proto-Anatolian and Proto-Indo-European share only one, possibly two words for an edible seed, neither of which distinguishes between a domesticated and a wild plant, and none at all for those early crops, lentils, peas and chickpeas.[9] On the other hand, Proto-Anatolian and Proto-Indo-European do share a rich vocabulary of herding, including words for 'horse', 'cow', 'sheep', 'yoke' and 'to pasture'. It looks more likely that the ancestral tongue was spoken by herders than farmers.

By way of a solution to this apparent discrepancy, it has been suggested that the Iranian farmers' descendants had lingered so long in the north, in areas unsuitable for cultivation, that they had become herders again and forgotten all their crop-growing knowledge. That is possible. But what about the westerly clustering of the Anatolian languages? Isn't it strange that they are absent in the east of the Turkish peninsula, if that is where they came from?

Not necessarily. For one thing, there *is* a possible trace of the first Anatolian-speakers' passage through those eastern parts. Around 2400 BCE, palace archivists at Ebla in Syria recorded about twenty names, including *Duduwashu* and *Aliwada*, that sound distinctly Anatolian to some linguists' ears. For another, archaeological evidence suggests that non-Indo-European-speaking groups in the Caucasus expanded at about the time that the ancestral tongue would have been fragmenting into its Indo-European and Anatolian daughters. Those non-Indo-Europeans could have pushed the first Anatolian-speakers gradually further west, until they ended up in the places where the Luwian, Lydian and Hittite kingdoms arose later. It is also possible that the first Anatolian-speakers reached the peninsula by boat from Yalta, Sochi or some other northern harbour – perhaps bringing silver to trade. The Roman historian Tacitus, writing around 100 CE, remarked that all ancient migrations came from the sea. He was probably thinking primarily of the Greeks who had colonised the Black Sea in the previous millennium, but his observation might hold for even earlier migrations.

Between 3000 BCE, when linguists say Proto-Anatolian was spoken, and the first written mention of the Hittites, a thousand years yawn blackly. A thousand years is a long time in human affairs; as a French historian once put it, it's the gap between Gaul and de Gaulle. But one could speculate that in that time, whether they came from east or west, by land or by sea, the first speakers of an Anatolian language dispersed to different parts of the peninsula and adapted to the farming life. There need not have been very many of them to begin with, and each pocket of Anatolian-speakers would have found itself surrounded by other farming groups who spoke non-Indo-European languages such as Hattian or Hurrian (a language recorded in northern Mesopotamia at about this time). Anatolian fragmented into its daughters, and from then on the fortunes of those daughters tracked the fortunes of their speakers.

When we catch our first glimpse of the Hittites in the archives, around 2000 BCE, their stronghold was the city of Kanesh, in the spectacular volcanic landscape of Cappadocia. They never actually called themselves 'Hittites'; that's a misnomer we owe to the Bible. They were the people of Kanesh, who spoke Neshili. To the southeast of Kanesh lay the Taurus Mountains and beyond the mountains lay Assyria. Donkey caravans now came winding through those mountains bearing Assyrian merchants and their wares. By 2000 BCE, bronze was being made by adding tin rather than toxic arsenic to copper, and the Assyrians brought copper which they hoped to exchange for tin. Soon they had established a colony at Kanesh and were trading in all kinds of luxury goods, including gold, silver, perfumes and woven textiles. To keep track of their transactions they introduced writing to the region, in the form of cuneiform. The first texts that refer to the Hittites were written by these merchants in their Semitic language, Old Assyrian. Besides Hittite personal names, they contain a sprinkling of Hittite loan-words including *ishpattallu* (night watch) and *ishiullu* (contract).

Thanks in large part to the Assyrians, Kanesh grew rich. But the Assyrian trade collapsed around 1800 BCE and a turbulent period ensued, in which neighbouring city-states were constantly at war. To the north-west of Kanesh, inside a large bend in the River Kızılırmak, lay the Hattian-controlled city of Hattush. Five days' gallop north of that, probably close to where the Kızılırmak tipped into the Black Sea, lay another powerful city called Zalpa where Hattian was also spoken. A Hittite king, Anittas, conquered Zalpa and Hattush and took control of the Kızılırmak Valley. Anittas favoured attacking his targets in the dark, which may be why the Assyrians of Kanesh felt the need to mount a night watch. The oldest surviving document written entirely in Hittite, in cuneiform, describes Anittas' conquest of Hattush in the first person: 'At night I took the city by force; I have sown weeds in its place. Should any king after me attempt to resettle Hattush, may the weather god of heaven strike him down.'

Eventually Hattushili I ascended to the throne. Apparently ignoring Anittas' curse, or feeling that it didn't apply to him, he moved the Hittite capital to Hattush. He tweaked the city's name to Hattusha, signalling continuity under the new regime, and changed his own name to the one we know: Hattushili means 'the one from Hattusha'. The royal archives were established under his reign, and they reveal that the Hittites co-opted the sophisticated political infrastructure that the Hattians had established, along with many Hattian gods and much of their culture. Among the Hattian loanwords that Hittite absorbed were a number that related to that older system of government: *labarna* (king), *tawannanna* (queen), *halmassuit* (throne), *halentiu* (palace) and *sahtarili* (singer-priest).

The two peoples fused, much as the Romans and Etruscans would do more than a thousand years later. Some scholars think of the Hittite founding myth, the one about the fertile Queen of Kanesh, as a retelling of the early history of the Hattians and

Hittites that glossed the violence and presented them as a family reunited. The queen gave birth to her thirty sons, and the myth went on: 'She said, "What a monster is this which I have borne?" She filled baskets with fat, put her sons in them, and launched them in the river. The river carried them to the sea to the land of Zalpa. But the gods took them up out of the sea and reared them.'[10]

The queen went on to have thirty daughters. This time she raised the royal offspring herself, at Kanesh, and when much later her estranged sons returned from Zalpa they met their sisters on the road without recognising them. Against the protestations of the youngest brother, the older boys slept with the older girls. To the eternal frustration of hittitologists, the clay tablet on which the myth is written breaks there and the end of the story is lost. They nevertheless speculate, based on similar myths recounted elsewhere, that the gods rewarded the youngest brother and sister for abstaining from incest, and that the two siblings went on to rule the joint kingdom wisely and well.

Hittite myth presented the two peoples as equal partners, in other words, but there are subtle linguistic clues that the Hittites retained the upper hand. Apart from the Hattian loans that their language absorbed, it changed very little. Hattian, on the other hand, was completely transformed by its entanglement with Hittite. In Hittite the verb was placed at the end of a sentence, in Hattian at the beginning. A comparison of texts written in the two languages shows that Hattian-speakers eventually switched to the Hittite word order, triggering a cascade of other changes in their language.

When you go to Boğazkale, as I did in the summer of 2023, you realise immediately why Hattushili wanted it for his capital. The ruins of the palace where he dictated his last words sit atop a basalt promontory that is protected on all sides by steep slopes or cliffs. Eagles nest in those cliffs, high above a tributary of the Kızılırmak that winds around their base. The lion king's successors built towards their cumulative vision of an imperial capital which, for as long as it thrived, dwarfed Troy and Athens. Exploiting the fact that water expands when it freezes, Hittite masons split pre-chiselled rock and flattened crags. With the blocks they hewed out of the crags they constructed ramparts that, in their scale and geometrical precision, rivalled the Egyptian pyramids. They sank rock cisterns to a depth of twenty metres (sixty-five feet) and cut near-vertical escape chutes into the palace's pedestal. The city had inner and outer defence walls, and the gates in the inner wall were shut and sealed each night. One was guarded by a pair of basalt sphinxes, another by lions. Their eyes are said to have flashed fire, perhaps because their now-hollow sockets contained red agates or jaspers. Hattusha was considered impregnable – until it wasn't.

Hattushili I expanded his domains southwards into what is now Syria. His adopted son Murshili I, whom he named his successor in his will ('The god will only install a lion in place of a lion'), pushed into Mesopotamia and sacked Babylon. The Hittites' fortunes waned after that, but in the fourteenth century BCE a canny ruler named Shuppililiuma I ascended the throne and restored them. For the next two centuries the Hittites boasted one of the most powerful empires in the Near East. Their modus operandi seems to have been to treat the subjugated peoples with benevolence, as long as they came quietly, but to show no mercy to those who resisted. The empire absorbed many elements of the cultures of its new subjects, such that the Hittites came to be known as the 'people of a thousand gods'. Some argue that Hattusha was cosmopolitan from as early as the reign of Hattushili I, with other Anatolian languages – notably Luwian and Palaic – being spoken within its walls.

In 1274 BCE, in what has sometimes been called the first world war because there had never yet been such an impressive turnout of infantry and cavalry, the Hittites and the Egyptians fought each other at Kadesh on the banks of the River Orontes, near the modern Syrian city of Homs. The Hittites brought their souped-up chariots (they had refined an earlier design by shifting the axle from the back to the middle), drawn by horses wearing feathered headdresses. The battle ended in a stalemate, with heavy casualties on both sides, though the Egyptian Pharaoh, Ramses II (Ramses 'the Great'), went home claiming victory: 'I slaughtered them. I killed them wherever they were.'

It took sixteen years, but Ramses eventually signed a peace treaty with the Hittite emperor Hattushili III and his queen Puduhepa. In a text inscribed on silver tablets, two of the world's superpowers promised to 'settle forever among them a good peace and a good fraternity'. The language of the treaty was Akkadian, the medium of international diplomacy at the time. The original tablets have been lost, but the text was engraved into the walls of Egyptian temples, and a Hittite version, written on clay, was stored in the archives at Hattusha. A replica of the Hittite version hangs in the United Nations headquarters in New York City, outside the Security Council chamber on the second floor.

Around 1200 BCE, not long after Puduhepa's death, Hattusha fell, the victim of a wave of collapse that reverberated around the Near East and brought the Bronze Age to a close. Traditionally this collapse was blamed on the Sea Peoples, ruthless pirates so named by the Egyptians, whose origins were veiled in mystery. Historical and scientific research has pieced together a more complex tableau. The trouble seems to have started with a drought that caused a severe and lasting famine in the region. A storm of earthquakes rippling along the faultlines that criss-cross the eastern Mediterranean forced starving, desperate migrants to take to the sea, and maritime trade stalled. Eventually, the staggeringly unequal civilisations of Egypt, Anatolia and Greece, heirs of the Uruk model of urbanism, faced plagues and peasant revolts that not even their towering ramparts could keep out. The Hittite capital had only ever been home to the royal family, bureaucrats, soldiers and priests. The peasants who supplied the food (including more than a hundred and eighty kinds of bread or bakery product, according to a museum in nearby Çorum) lived beyond the walls, shut out from the machinery of state and the wealth that oiled it.

Hattusha burned and the Hittite language, soon to pass into oblivion, was baked hard into the clay. Some claim that it was already dead by the time of the crisis, having been replaced as the city's everyday language by Luwian. Luwian survived for another five hundred years, while Lydian was still being spoken a few centuries before the birth of Christ. At some point in the first millennium of our era, however, the last Anatolian language fell silent, never to be heard again.

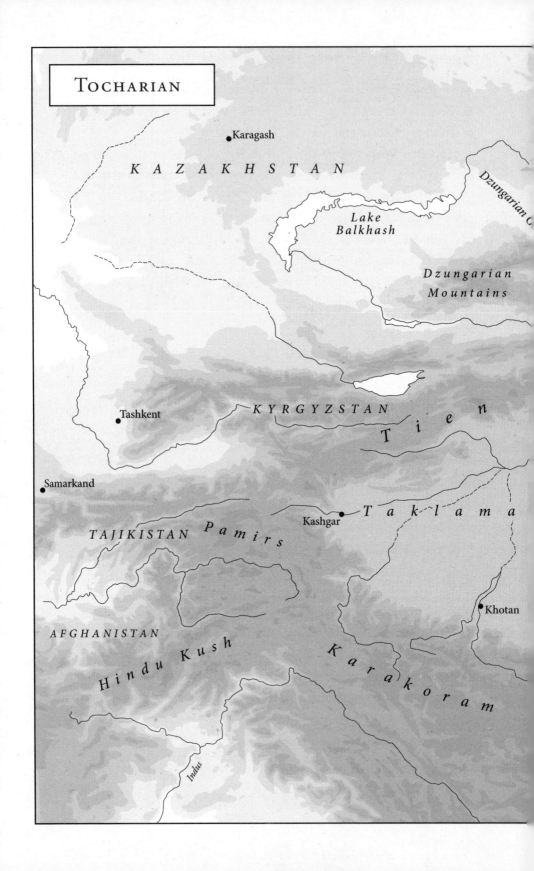

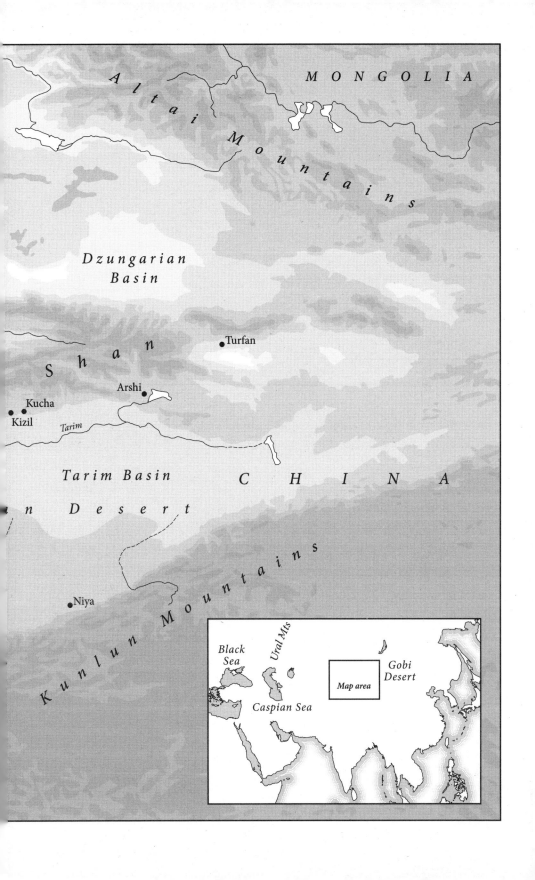

4

Over the Range

Tocharian

The wind blew, the sand ran, and the dunes dissolved, only to reform before the visitors' eyes. Their camels plodded on through that immense hourglass, until it occurred to them that the streaming planes were no longer smooth. Here protruded a wooden post, there a bleached tree trunk. Dismounting to explore further, they discovered more dead trees – mulberries, of all things, and curiously aligned. Eventually they traced out avenues and irrigation channels, pools and gardens. Then they took their spades and dug. From the drifts emerged fireplaces and eating trays; boot lasts and mousetraps; and finally, the pages of manuscripts, excellently preserved by the dry desert climate.

It was the autumn of 1906, and the meticulous, Hungarian-born archaeologist Marc Aurel Stein had just discovered Niya, one of the oasis cities that once encircled the Taklamakan Desert like a string of mala beads. It was his second expedition to what is now Xinjiang, north-west China, and he would make two more, always accompanied by his faithful terrier Dash (seven terriers, all Dash).[1] Besides the world's oldest printed text, the Diamond Sutra, and the Jade Gate – the frontier post where caravans once entered

China – Stein would discover much of the Silk Roads on behalf of his adopted country, Britain.

The last decade of the nineteenth century and the first of the twentieth were a time of intense international interest in eastern Central Asia. The Swedes had been the first to explore Xinjiang, with the Chinese, but the British, Germans, French and Russians weren't far behind, nor was the Japanese abbot Count Ōtani Kōzui. The ambitions of the governments that funded these expeditions were more or less nakedly imperialistic, and the methods by which the explorers appropriated antiquities would be considered unethical today. Nevertheless their discoveries lifted a veil on the region's storied past.

Taklamakan is usually translated as 'you go in but you never come out', though experts in the Uyghur language that has been spoken in Xinjiang for more than a thousand years say that it means something closer to 'abandoned place'. The desert lies in a vast depression, the Tarim Basin, that millions of years ago contained a sea. On three sides it is enclosed by mountains and on the fourth by the Gobi Desert. Lying south of Siberia and north of Tibet, it couldn't be more landlocked. It must also rank among the places on Earth that are least fit for human habitation. Starting about four thousand years ago, however, humans learned to capture the water that ran off the mountains before it dissipated into the sand, and used it to irrigate their crops. By the first millennium BCE cities had arisen from those oases that served as vital staging posts on the Silk Roads. As long as they could hold their own against the nomads to the north and the Chinese to the east, those watering holes thrived. But starting in the ninth century CE, they entered an inexorable decline. The people left, the sand encroached, and eventually explorers came in search of the buried cities of legend.

By 1906 Stein had already excavated the largest oasis on the desert's southern rim, Khotan. Others had begun to disinter the cities of the northern rim, including Kucha and Turfan further east. Occasionally the explorers stumbled on human remains, mummified naturally by the fierce Tarim winters. They noted their Western appearance and moved on, being more interested in the manuscripts which they carried away by the armful from caves that had once served as libraries. Sometimes pages had blown out of the libraries, and these they plucked from the sand. The caves were decorated with exquisite murals from which they cut sections. At a cave complex called Kizil, the murals depicted scenes from the Buddha's life along with portraits of the benefactors presenting gifts. The benefactors were portrayed as warriors or monks, with *tikka* marks on their foreheads and colourful silken robes, but they had reddish hair and green eyes. To the foreigners, they too looked curiously Western.

Most of the manuscripts vanished into European museums – in Berlin, Paris, London and Saint Petersburg – where philologists pored over them in quiet back-rooms. They were written in many different languages, some of which were known, some unknown. The Indian script most often used, Brahmi, was one the philologists were familiar with, since it had been deciphered in the previous century.[2] And because the manuscripts included translations of well-known Buddhist texts, they were able to decode the unknown languages relatively easily. In 1908 two German scholars, Emil Sieg and Wilhelm Siegling, announced that one of the previously unknown languages recorded in the Berlin collection was Indo-European. They called it 'Tocharian' because they had seen it referred to in Old Uyghur documents as *tohri* and they assumed, wrongly as it transpired, that this was a reference to the medieval region of Tokharistan, far to the west of Tarim.[3] (The Tocharians named themselves after some of their towns, being the people of Arshi or Kucha.) The misnomer stuck and in time Sieg and Siegling were credited with the discovery of Proto-Indo-European's second eldest daughter. Their peers could hardly believe it: an Indo-European language was once spoken in the Far East! Rumours began to swirl of red-haired Europeans journeying east long before Marco Polo, and depositing their language at the doorstep of China.

If you were to pool all the Tocharian texts that the former imperial powers divided up between them, you'd count about ten thousand of them. They capture the glory days of the Tocharian language and civilisation, between the fifth and tenth centuries CE. Most are mere fragments, pages torn from religious texts, dramas and treatises on magic and medicine. There is a version of the Pygmalion myth, in which a painter falls in love with a mechanical girl made of wood, and a Buddhist tale about a white elephant with six tusks who completes the path to enlightenment after his jealous wife has him killed. And there is a love poem, which begins thus:

> *Mā ñi cisa noṣ śomo ñem wnolme lāre tāka,*
> *Mā ra postaṃ cisa lāre mäsketär-ñ.*
> *Ciṣṣe laraumñe ciṣṣe ārtañye pelke kalttarr śolämpa ṣṣe.*

> Earlier there was no person dearer to me than you,
> And later too there was none dearer.
> The love for you, the delight in you is breath together with life.[4]

These shards sparkle, but they shed almost no light on how the Tocharians lived. Besides a few caravan passes, notes dealing with the running of monasteries and scraps of graffiti, no record has survived of the more mundane aspects of their existence. James Mallory has said that trying to reconstruct the Tocharian world from the surviving texts would be like trying to reconstruct life in the late Roman Empire from the Christian liturgy. Fortunately we have other sources of information about the Tocharians. There are the cave paintings in which they depict themselves. And there are the descriptions left by their neighbours to the east, the Chinese of the Han and later dynasties.

Until about the turn of the Common Era, the Chinese regarded the Tarim Basin as strictly beyond the pale; a place of fabulous humans and beasts. As they were drawn into closer dealings with it they recorded their impressions. Their chronicles cover a period of about five hundred years when the cities were thriving, having grown rich on China's trade with the West as well as their own agriculture and minerals. Kucha, the largest of the oases, was home to about a hundred thousand souls. Inside its thick walls, besides residential and administrative quarters, there were hundreds of Buddhist temples and *stupas* or shrines. The king lived in a palace that shimmered 'like an abode of the gods'.[5] Khotan, on the southern route, was the centre of Tarim's silk industry. Legend had it that when the Chinese banned the export of silkworms, mulberry trees (whose leaves silkworms eat) and all knowledge of the silk-making process, a canny king of Khotan asked for a Chinese princess's hand in marriage. He told her that if she wished to be kept in silk she would have to smuggle silkworms and mulberry seeds out in her headdress, which she duly did.

The Chinese regarded the Tocharians as heavy-drinking, decadent barbarians. Kucha was famous for its music and dancing girls, not to mention its wine cellars and the flock of a thousand peacocks upon which its nobles liked to feast. The Chinese portrait provides an earthy, epicurean counterpoise to the Tocharians' depiction of their spiritual life. They were early converts to Buddhism, after it spread beyond India and Nepal, and having entered into commerce with their eastern neighbour they played an important part in funnelling that religion into China. The monks of Kucha translated Buddhist texts from Indian languages into local ones, including Tocharian, and word spread along the trade routes. By the fourth century CE there were sacred Buddhist sites all over north-west China, and a hundred years later Buddhism was competing with Confucianism in that country. The Uyghurs, whose Turkic language came to dominate in Xinjiang from the ninth century CE, were disciples of the Tocharians before they converted to Islam starting in the following century.

In identifying Tocharian as Indo-European, Sieg and Siegling were essentially announcing that its origins lay thousands of kilometres to the west, beyond the Urals. At some point, an ancestral tongue must have crossed Eurasia to reach China. For over a hundred years, scholars have tried to understand who brought that ancestral tongue, when, by what route, and perhaps most intriguingly, why. Beginning with the study of the language itself, they have drawn on archaeology and genetics to build a picture of the speakers of Tocharian before their language was written down. That picture is very much a work in progress, but it is one that has filled out significantly in the last decade.

As soon as they began deciphering Tocharian, the German philologists realised that it was not one but two languages. Since the two were closely related, they labelled them Tocharian A and Tocharian B.[6] Tocharian A was probably dying out by the seventh century CE, since its use was already restricted to the religious domain, but B remained vigorous until at least the tenth century. B was spoken mainly in the western part of the basin, including Kucha, while A was spoken further east, and when they were first written down they were about as different as Italian and Romanian. Based on those differences, linguists estimate that they split from a common ancestor, Proto-Tocharian, in the early first millennium BCE. In the early first millennium BCE, the cities of the Tarim Basin weren't yet cities. The Buddha wasn't yet born. Whoever the speakers of Tocharian were at that time, they were not the proselytising urbanites they would become. So who were they, and *where* were they?

The core vocabulary of Tocharian was clearly inherited from Proto-Indo-European. You can see that in these Tocharian B – Latin – English triplets: *pācer* – *pater* – *father*, *mācer* – *māter* – *mother*, *procer* – *frāter* – *brother* and *ṣer* – *soror* – *sister*. Or in the Tocharian B words for 'cow' (*keu*), 'ox' (*okso*) and 'to milk' (*mälk*). One other branch of the Indo-European family headed east into greater Asia, and that was Indo-Iranian (the subject of Chapter Six). It would be tempting to think that Tocharian had left with Indo-Iranian, even that they had started out as one language and split en route, but linguists considered that unlikely from the start. There are important differences between Tocharian and Indo-Iranian that speak to a lengthy separation in time.

The most important of those differences is a sound change called satemisation. The Indo-European family falls out into centum and satem languages, depending on whether words like the one for a 'hundred' (*centum* in Latin, *satəm* in Avestan) are pronounced with a hard *k* or a soft *s* sound at the front. Italic, Celtic and Germanic are all centum languages, while Indic and Iranian are satem languages, and counter-intuitively – given that it was a language of the East – Tocharian belongs with the first group (the Tocharian A and B words for a hundred were *känt* and *kante* respectively). The shift from *k* to *s* occurred after Tocharian and the western branches had departed the Indo-European homeland, leaving them unaffected by it.

Tocharian left earlier than Indo-Iranian, then. But linguists think that it left after Anatolian split from Proto-Indo-European, because unlike Anatolian it inherited the Proto-Indo-European wheel word, *$^*k^wek^wlos$*. This became *kokale* in Tocharian B and *kukal* in Tocharian A, both of which meant 'wagon' rather than 'wheel' – a case of semantic shift that is exactly what linguists would expect after millennia of linguistic and technological evolution (remember that the oldest Tocharian documents were written three thousand years after Proto-Indo-European expired). Conventionally, Tocharian has been regarded as the second language to leave the Proto-Indo-European homeland after Anatolian (although if you subscribe to the theory that Anatolian and Proto-Indo-European are sisters, Tocharian finds itself promoted to first daughter status). But Tocharian also holds clues as to where it went *after* leaving that homeland. As with Romani, which departed India heading west, some linguists claim that they can reconstruct its itinerary through the influences that it absorbed en route.

As befits a language of the Silk Roads, Tocharian was intrinsically cosmopolitan. It borrowed not only words, but also sounds and grammatical structures, and from quite a variety of languages. At some point in its journey from the Indo-European homeland towards the Tarim Basin, it underwent a massive grammatical and phonological rupture. It lost all its voiced stop consonants, such that its speakers no longer produced (or heard) the difference between *k* and *g*, or between *p* and *b*. It also jettisoned most of the Proto-Indo-European system for marking case in nouns (the system which, at its simplest, indicates whether a noun is the object or the subject of a verb) and replaced it with a new one. It underwent many other changes besides, and taken together these point strongly to contact with Uralic – the language spoken by the hunter-gatherers of the northern Eurasian forests – and specifically with its eastern branch, Samoyedic.[7] Michaël Peyrot, a linguist at Leiden University, says that it's as if some early Samoyedic-speakers had taken up an Indo-European language. They spoke it with such a strong accent, and so often violated its rules, that they ended up changing it in profound ways. Today, the Samoyedic languages are restricted to northern Siberia and the Russian Arctic, but they were probably once spoken further south – as far south as the Altai.

We may have evidence for an Indo-European language lingering in a mountain range north of the Tarim Basin, then. But we also have clues as to where it went from there, because the language that emerged from the entanglement with Uralic eventually came into contact with the Iranic languages. The Iranic languages, as we've said, left the Indo-European homeland later than Tocharian and may have taken a different path east. By the time the two branches met again they were obeying different sound laws (including satemisation, in the case of Iranic), and the loanwords they exchanged therefore bear the stamp of those laws. As with the Persian loans into Romani, those words help linguists work out where and when the encounter, or encounters, took place.

The earliest Iranic loans to Tocharian are from Old Iranian, the language of the horse-riding nomads of Central Asia in the Bronze and Iron Ages. Those nomads inhabited a vast territory, but in the east they were hemmed in by an inflexible boundary – the chain of mountains, thousands of kilometres long, that separates the western steppe from the eastern steppe and China. Around 1200 BCE, when a new spell of climate change caused that part of the world to grow wetter, the grasslands to the east of the mountains expanded, giving the nomads an incentive to cross to the other side. The intrepid ones who succeeded would have come through the only break in the wall, the Dzungarian Gate – the famed pass that Herodotus credited with being the source of the north wind. The Dzungarian Gate leads down into the Dzungarian Basin, another depression sandwiched between the Altai to the north and Tarim to the south. Archaeologists have detected vestiges of the nomads in Dzungaria between the thirteenth and ninth centuries BCE, when it would have been a lush haven for herders. It is there that they might have encountered the speakers of a pre-Tocharian language who had migrated south from the Altai. One relic of that encounter was the Old Iranian word *zaēna* (weapon), which became *tsain* (arrow) in Tocharian B.

The Middle Iranian languages, which succeeded Old Iranian, also loaned words to Tocharian. Since they included Sogdian, one of the languages of trade on the Silk Roads, and Khotanese, the language of the silk city Khotan, we may infer that Tocharian had reached the Tarim Basin by the time it crossed paths with them, in the last millennium BCE.[8] (From a Sogdian word meaning 'wine', itself derived from Proto-Indo-European *medhu* or 'mead', Tocharian B took its word for 'alcohol', *mot*.)

Later still, Tocharian brushed with Sanskrit and its Indic relatives. Many of the Sanskrit words it received have a distinctly Buddhist flavour, meaning that they date themselves to the last few centuries BCE. The Tocharian B word for the enlightened being called a 'bodhisattva', for example, was *bodhisatve*. Linguists know that this was a loan from Sanskrit and not inherited from the two languages' shared ancestor, because Tocharian did not contain the sounds *d* or *dh*. These were among the voiced stop consonants that it purged in its earlier clash with Uralic, and Tocharians would have pronounced both as *t*.

But Tocharian wasn't just a sponge for other languages; it also lent words. The arm's-length relationship between the Chinese and their barbarian neighbours to the west translates into a striking impermeability of the two languages to each other, but there were exceptions. Tocharian lent a word for 'honey' to Old Chinese (the ancestor of Modern Chinese *mì*). It is even possible that Tocharian lent a word to English – a tribute to the long tentacles of the Silk Roads. As far-fetched as it may seem, literally, linguists consider it more likely that a Sanskrit word meaning 'monk' (*śramaṇas*) passed into Tocharian B (becoming *ṣamāne*) before entering English as 'shaman', than that English took the word directly from Sanskrit.†

Linguists quibble about the provenance of these loans, but for some they present convincing evidence of the path that Tocharian took to become itself: from the Pontic-Caspian steppe east to the Altai, from the Altai south to the Dzungarian Basin, and from Dzungaria south to Tarim. It's quite a leap to connect the Yamnaya

---

† *ś* in Sanskrit and *ṣ* in Tocharian B both represent the sound English-speakers write as *sh*.

of what are now Ukraine and Russia with the Tocharian monks and troubadours on linguistic sleuthing alone, however, and even those scholars who find the theory persuasive would like more evidence that prehistoric people actually travelled that route, carrying an Indo-European language with them. The spotlight therefore turns to archaeology, and genetics.

It was in the 1980s that word of the Afanasievo culture first reached Western archaeologists, though those who read Russian might have learned about it earlier. The carriers of this culture lived in the Altai Mountains and the Minusinsk Basin of Siberia at about the same time as the Yamnaya were a presence further west.[9] They had much in common with the Yamnaya. They were herders, horse-breeders and metalworkers, and they buried their dead in a similar way: laying them on their backs with their knees raised, sprinkling them with red ochre and covering them with a kurgan. Their way of life bore little resemblance to that of the hunter-gatherers who had long inhabited the Altai, and there were even suggestions, based on examinations of their skulls, that they were European in appearance. Archaeologists concluded that the ancestors of the Afanasievo had been intruders in the region, possibly migrants from the west.

That hypothesis was confirmed in 2015, when geneticists from the world's third great centre for palaeogenetics, at the University of Copenhagen in Denmark, announced that the people of the Yamnaya and Afanasievo cultures were genetically almost identical. It was a watershed moment in the Tocharian story, because the only viable explanation seemed to be that a group of Yamnaya had migrated east to the Altai, carrying their genes, their language and their lifestyle with them. The people of the Afanasievo culture were those migrants' descendants. Then hard on the heels of that stunning discovery came another.

Just over halfway between the Ural River and the Altai, at Karagash in what is now central Kazakhstan, an extinct volcano rises from the steppe. Inside its crater lies a forested plain, and on its exterior slopes stands a kurgan that, when it was freshly built, would have been visible for miles around: an unmissable signpost. In 2018, the Danish group reported that the individual buried beneath that kurgan shared the genetic make-up of the Yamnaya and the Afanasievo. They appeared to have identified a stopover on the migrants' route.

These findings convinced many that the Yamnaya had carried their language across Eurasia, bequeathing it to their Afanasievo descendants. But did that language become Tocharian? The Afanasievo disappeared from the archaeological record around 2500 BCE. The oldest surviving documents in Tocharian don't turn up for another thirty centuries – quite the hiatus. Ancient Egypt rose and fell within thirty centuries. Many other scenarios could have played out in that civilisation-guzzling interval. It seems hard to dispute that a group of Yamnaya went east, but even if they transmitted their language to the Afanasievo, that language could have died out with the Afanasievo. Thousands of years might then have elapsed in which no Indo-European language was uttered in the vicinity of China, until a later Indo-European language arrived and seeded Tocharian. Some scholars continued to argue, as they had before the discovery of the Yamnaya-Afanasievo genetic link, that farmers from what is now Afghanistan had brought the ancestor of Tocharian around 2000 BCE, along with the irrigation techniques that allowed humans to settle Tarim. Others claimed that Tocharian's ancestor had diffused through the mountain corridor that links the Iranian Plateau to the Altai. The herding communities of those mountains were known to have traded knowledge and mates through long-distance exchange networks, and perhaps a language had come along for the ride. All three scenarios were plausible, and the only way to boost the odds in favour of the Afanasievo, as intermediaries, was to show that their genetic ancestry had found its way to Tarim. Genes and language don't always travel together, but if humans had migrated south from the Altai, it was certainly possible that they had taken an Indo-European language with them.

In the last few years geneticists have indeed found evidence for such a migration, or migrations. They have shown that populations who succeeded the Afanasievo in the Altai Mountains inherited both their ancestry and their culture, and they have found the same ancestry further south, in the Dzungarian Basin. The final – but still missing – link in the chain was that between Dzungaria and Tarim, and, naturally enough, attention turned to the mummies that Stein had photographed in the early twentieth century. ('It was a strange sensation', he wrote later, 'to look down on figures which but for the parched skin seemed like that of men asleep.')

Hundreds of mummies would be excavated before the twentieth century was out. They shocked and continue to shock almost everyone who saw them, partly because of how unskeletal they appeared, after four thousand years, and partly because of their, for some, exotic appearance. They were typically tall, with fair hair, pale skin, long noses and deep-set eyes. Similarities were noted between their clothing and that of the Tocharian knights portrayed in the cave paintings at Kizil. Some of the younger ones, dated to around 1000 BCE, had gone to their graves wrapped in woollen textiles with tartan-like designs, reminiscent of Celtic plaids being woven in Europe at that time. Many speculated that they were the descendants of steppe nomads from the west, and that they had spoken Proto-Tocharian.

In 2021, scientists from China and Germany put the speculation to rest. Having analysed the genomes of thirteen of the oldest mummies, they reported that those individuals were a genetic vestige of the hunter-gatherers who had inhabited the eastern end of Eurasia since before the last ice age. They had not interbred with any of the populations around them – not the Afanasievo, nor the Afghan farmers, nor the herders of the mountain corridor. If they looked European, it was more likely to be because their ancestors had bequeathed genes shaping skin and hair colour to ancient Europeans, than because Europeans had come east. A century's worth of European stereotyping evaporated into thin air: they were as local as it was possible to be.

The mummies were among the first human beings to make their home at the edge of the Taklamakan Desert. If they were able to do so, it was because of their willingness to borrow from cultures around them. They herded cattle and made cheese, a skill they may well have learned from descendants of the Afanasievo, who were themselves descendants of the Yamnaya. But they also grew millet and irrigated their crops, and those were eastern innovations. Cowrie shells have been found in their graves that can only have come from a very long way away. As for the tartan, they may have invented it or imitated an invention of others, one they had learned about through far-reaching networks of communication and exchange.

The Tarim mummies are a reminder of the complex links between genes, culture and language. As Mallory has observed, they blended eastern and western cultures two thousand years before East and West officially acknowledged each other's existence, and yet they remained genetically 'pure'; an isolate. As for what language they spoke, it's anybody's guess. It could have been one related to that of their hunter-gatherer forebears, or one they borrowed from the cheese-makers, or the millet-growers, or the tartan-weavers, or another one again. It is possible that other people brought steppe ancestry to the Tarim Basin, once the mummies were already established there, but if so that ancestry has yet to be found. The search continues, and in the meantime nobody can definitively say how Tocharian got to China. Ask Michaël Peyrot how he thinks it happened, and he will tell you a story that can be paraphrased thus:

Migrants from the Altai arrived in the Dzungarian Basin before 2500 BCE, carrying the genes of the Yamnaya and the Afanasievo and speaking a forerunner of Tocharian. In Dzungaria they encountered ancestors of the Tarim mummy population, whom they displaced to the south, but not before learning from them the techniques of irrigation agriculture and the vocabulary to go with it (and sharing the secrets of cheese-making in return). Around 1200 BCE, mounted steppe warriors came thundering down from the Dzungarian Gate, perhaps in a blizzard of arrows, and pushed the Proto-Tocharians south in turn. When the latter reached the Taklamakan Desert they split into two streams to skirt it, and their language split too, into A and B. They built the cities of the Tarim oases, which later received waves of Iranic-speaking migrants from the south and west. The trade routes that were operating by 1000 BCE began to deal in silk from 300 BCE, and in enlightenment not long after that. A golden era ensued, which was captured in cave paintings that are now scattered between the museums of former imperial powers. By the ninth century CE the Tocharians were losing the power struggle with the Chinese to the east and the Turkic-speakers to the north, and by the tenth century Tocharian was extinct, or nearly.

Harald Ringbauer's pulse raced. He checked the plot again, but there was no mistake: the two individuals were related. They could have been second cousins, he calculated, or first cousins once removed. Both had died about five thousand years ago. And this was the part that made him check the computer's readout yet again, even though he knew that the quality of the DNA had been excellent and there was no reasonable chance of error: one of the cousins had been buried in the Don Valley, north-east of Rostov in Russia, the other more than three thousand kilometres (two thousand miles) to the east, in the Altai.[10]

It was 2020, the lull following the first surge of the Covid-19 pandemic. A postdoctoral fellow with David Reich's group at Harvard, Ringbauer had managed to keep working while observing strict distancing rules. Not far away, shielding in his own office, was mathematician Nick Patterson. The septuagenarian former chess champion and code-breaker had decided at an age when others are comfortably retired that there were still interesting problems for him to solve, and that the most interesting of all lay in biology. Since joining Reich's group as the 'data guy', he had not been disappointed. Now he was looking at a similar plot to Ringbauer and having a similarly hard time believing his eyes, but he came to the same conclusion as his younger colleague: as fantastical as it seemed, a group of Yamnaya had travelled the length of the Eurasian steppe in a few decades, perhaps less.

Ever since the Copenhagen group had shown that the people of the Afanasievo and Yamnaya cultures were related, archaeologists had acknowledged that there had been a migration across the steppe, but they tended to think of it as an expansion similar to that of the Neolithic farmers of Anatolia: a wave of advance inching eastwards over hundreds of years and many generations. What Ringbauer and Patterson were seeing wasn't a migration in that sense; it was a *trek*. Epic treks are known about from history: Mormons hit the wagon trail in Illinois in the nineteenth century, heading for Utah; Norsemen braved the stormy North Atlantic in the ninth century, to settle Iceland. This one dwarfed both of those, and it had happened thousands of years earlier. Patterson realised that the event that he and Ringbauer had detected was, in his words, 'almost unique in human history'.

Once a migration route has been established, people typically move back and forth along it, and that is likely to have happened in this case too. But the researchers' discovery of relatives buried at either end indicated that at least one set of migrants, probably the pioneers, had gone fast. Exactly how fast is hard to say. The cousins, several pairs of whom have been identified by now, could have died within a few years or a few generations of each other. Radiocarbon dating can't offer more precision than that, for now. Some archaeologists think that a hundred years would be a reasonable estimate for the duration of the trek, although they don't rule out a decade. David Anthony, ever the bold thinker, has gone further than most, proposing that the first group crossed in as little as two years. He thinks that the forested crater at Karagash may have been their only substantial stopover; the place where they passed the one intervening winter before moving on again the following spring. If he is right about the speed of the initial crossing, then that only adds urgency to the question: why?

'It always felt somehow as if the land had eyes,' said Australian explorer Tim Cope about the expedition that he embarked upon in 2004, retracing Genghis Khan's journey across the steppe. During the more than three years that it took him to ride from Mongolia to Hungary, he confronted horse thieves, dog thieves, illness (his own and his animals'), hunger, failing equipment, bad weather and wolves. The wolves' night-time howling made him fear for his horses more than for himself, he said, because without his horses he would have died. He came to understand why the ancients were so often buried with their mounts: 'The horse literally carries people to life in the steppe, through life, and then, they hope, off into the afterlife.'[11]

Five thousand years ago, travellers would have faced many of the same challenges as Cope, without any of the modern conveniences. Even then, the steppe would have been occupied by potentially hostile groups, including the horse-rearing Botai. If the natives weren't out in the open, they were certainly in the river valleys, and the Yamnaya would have had to cross their territory.

Temperatures can soar to forty Celsius (a hundred and four degrees Fahrenheit) in summer, and drop to minus forty (minus forty Fahrenheit) in winter. It's easy to die of exposure or starvation. Cope was young, fit and alone, and he had GPS. The Yamnaya would have brought herds, and among their number were probably pregnant women, infants and elderly and sick people. They were almost certainly preceded by scouts, so they knew where they were going, and they may have had outriders. But most of the group would have proceeded on foot beside the wagons, across roadless plains and at an ox's stately pace. They undoubtedly risked their lives, but this was something other than a sacred spring. Most experts agree that a very exceptional set of circumstances must have prompted such an undertaking.

We can't ask them what it was, but that doesn't mean there is nothing we can say about their motives. For example, climate change may have been a factor, but if it was it's unlikely to have been the only one, because climate varies greatly across the steppe. Even if they had been suffering a drought west of the Urals, there's a good chance that simply crossing the mountains would have solved their problems, landing them in grass higher than their heads. They wouldn't have needed to go as far as they did. Michael Frachetti, an archaeologist who specialises in steppe societies, suggests that the push factor might have been ideological: the tyranny of small differences. Burial rites are considered a powerful reflection of a people's worldview, and even the earliest Afanasievo graves differ slightly from those of the Yamnaya. Unlike the Yamnaya, the Afanasievo placed stone enclosures around their kurgans.

Frachetti wonders if a splinter group of Yamnaya set off to establish a new religion, or to return to a 'purer' state by reviving an old one. One thing we can't see before the invention of writing is the charismatic leader who persuades his or her followers to break with tradition and build a new kind of society, but charismatic leaders must always have existed. Reasons to leave may accumulate, but it often takes an individual who inspires confidence, someone who can create a sense of unity and purpose, to translate those reasons into action. We could be looking at the signature of such a person in this trek. Did a prehistoric prophet lead some idealists into the unknown? We will never know.

And if three thousand kilometres is an awfully long way to go to find grass, it's also an awfully long way to go to escape your family. It's likely there were pull factors too. Altai is a Turkic word meaning 'gold', and besides gold there is copper and tin in those mountains – the two ingredients of bronze. The Yamnaya were skilled metallurgists. Their scouts could have brought them word of untapped seams in the East. Whatever the draw was, it was powerful enough to entice those plain-dwellers into the hills. Once they got there, they would have found reasons to stay. Frachetti has shown that the semi-alpine grasses on the Altai's higher slopes are extremely nutritious. Animals feeding on them in summer would quickly have grown fat. 'Pastoralists would notice that and consider it magic,' says Anthony, especially since up there, they were closer to the gods.

The Afanasievo did indeed go high. Their graves have been found on the Ukok Plateau at over two thousand metres' (seven thousand feet) altitude. Ukok, home to the steppe eagle and the snow leopard, is a remnant of what's known as mammoth steppe – essentially, the steppe as it looked in the last ice age. These cold, dry grasslands once connected the planet's northern realms, creating natural migration corridors for animals that needed to move to survive fluctuating weather patterns. The mammoth steppe ceased to exist in a contiguous sense after that ice age, and many of its indigenous species vanished. Five thousand years ago, however, mammoths still survived in pockets of it. Anthony wonders if a scout could have spied one on the plateau, munching the flowers that they loved to eat, and given such a vivid, breathless account of it to his clan that the others had to see it for themselves.

We are in the realm of speculation now. There is no evidence that mammoths roamed the Altai at that time, or that the Yamnaya/Afanasievo encountered them, but it is not out of the question. That extra incentive might not even have been needed. The Altai are beautiful. Prehistoric people would have been as sensitive to natural beauty as we are, perhaps even more so since they invested the physical world with supernatural powers. And paradise is something that humans have yearned for ever since they left the first one.

When Frachetti is in the field in Kazakhstan, he sometimes gets invited into a nomad's yurt for refreshments. He has been treated to steppe hospitality often enough over the years that he has detected a recurring theme in his hosts' interiors. There, in the world's largest landlocked country, he finds himself gazing at images of idyllic beaches in the Maldives or some other impossibly distant place. The yearning for a better world is alive and well and as doomed to disappointment as it ever was (the word 'utopia' contains that disappointment within it, since it means 'nowhere'). It's impossible to know the Yamnaya's intentions or desires, but both Frachetti and Anthony suspect that they might have been seeking their own El Dorado. If so, their quest carried them beyond the north wind, and Tocharian was its legacy.

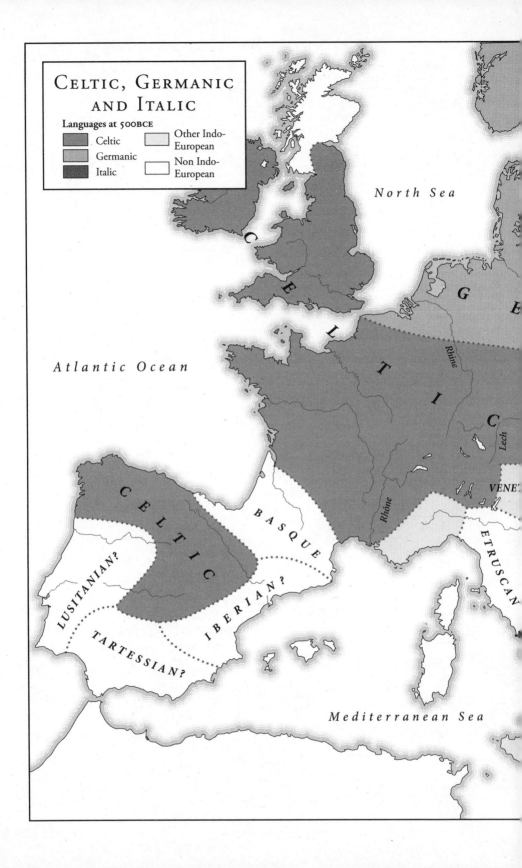

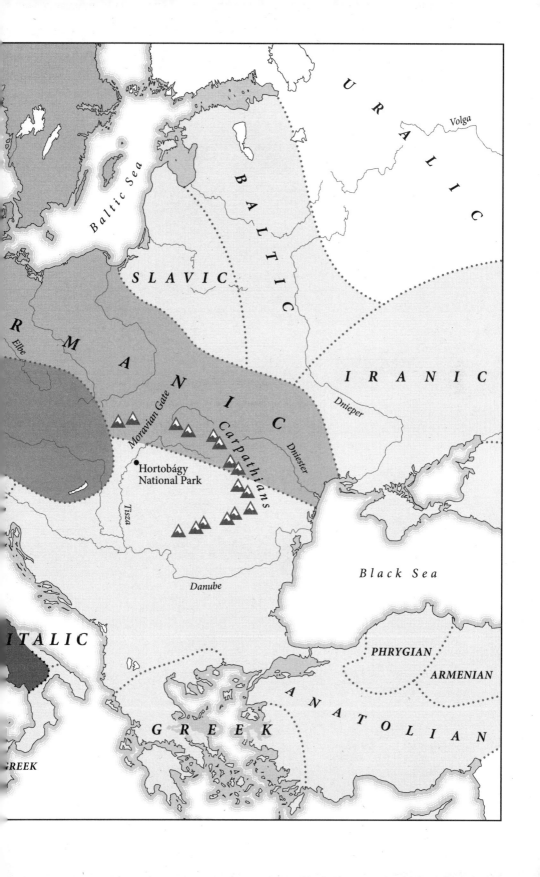

5

Lark Rising

Celtic, Germanic, Italic

History in western Europe ignites as a spark at the tip of Italy in the eighth century BCE. The Greeks, who had been writing in one script or another for seven hundred years, landed in Sicily and the southern part of the boot to trade and settle, bringing the alphabet with them. The Etruscans further north borrowed it, with modifications, and their Italic-speaking neighbours borrowed theirs. From Italy, writing spread north and west in a travelling wave. If you think of the linguistic landscape of Europe as a skein of invisible ink, writing was the candle flame that illuminated it. And what that flame captured, in the middle of the first millennium BCE – around the time that the Parthenon was being erected in Athens – was a sooty mosaic of Indo-European and non-Indo-European languages.

Celtic was spoken across a broad swathe of the continent from Spain to Bulgaria. To the north of the Celts, though still in preliterate darkness for now, lay the Germanic languages. Italic was spoken in Italy, but the Roman Republic was still in its infancy and Latin was not yet the all-conquering lingua franca that it would become. It shared the peninsula with other Italic languages, as well as with Celtic in the north, Greek in the south and at least one non-Indo-European language, Etruscan. Relatives of Etruscan were spoken alongside Celtic in the Alps and further east, on the Aegean island of Lemnos. Basque was dug into its foxhole between the mountains and the deep blue sea. Hungarian wasn't even a glimmer in a Magyar's eye.

To be sure, history had a faltering start in Europe. The first writings that were preserved were mainly inscriptions in stone. Sometimes they consisted of a single word, usually a name. Seven words was long. Writing winked in and out. There were periods, and peoples, who left us nothing written at all. Like the Greeks, the Etruscans and Italic-speakers wrote from right to left at first. Later they went through a phase called *boustrophedon* or 'ox-turning', when a line written right to left alternated with one written left to right, until they plumped definitively for left to right.

The Celts were next after the Italic-speakers to take up the alphabet, in the region around the northern Italian lakes and the plain of the River Po. The first Celtic inscriptions, which exist in a few different scripts, date to the sixth century BCE. By the time the Celtic-speaking Britons got around to writing, the Romans had arrived on their shores, and they recorded their language in the alphabet of the interlopers. The Irish invented their own alphabet around the fourth century CE. Called Ogam and consisting of strokes and notches carved into the edges of standing stones, some scholars consider it a work of genius. Unlike many other early European scripts, Ogam matches letters to sounds exceptionally well. Its anonymous inventor, probably not a monk since it predated the arrival of Christianity in Ireland, must have conducted a rigorous phonological analysis of the language that he or she wished to capture.

The Germani came relatively late to writing. Like the Irish, they must have been aware of the Mediterranean alphabets, but they were far enough from Italy that their response to them was highly original. They came up with the runes. Some say that the runic alphabet grew out of the Latin one, others that its model was Etruscan. The first uncontested runic inscriptions date to the second century CE and were found in Jutland, the peninsula that today straddles Denmark and Germany.

Arguably no other writing system besides the Egyptian hieroglyphs is wrapped in such an aura of mysticism as the runes (the word actually means 'secret'), though writing must have struck all those who encountered it for the first time as magical.[1] Imagine being shown a scratch on a mossy log and told that it's your 'name'. It has no head, arms or legs, yet others who see it will think only of you. When you consider all that writing made possible, from seducing and defrauding people over long distances, to keeping track of complex transactions and claiming property and kinship after your death, it's not hard to believe that people felt reverence for it; that in their minds it could alter a person's destiny. It would have inspired awe and even fear, like early photography. The written word retained that talismanic power for a very long time, as attested by the *Poetic Edda*.

The *Poetic Edda* is a collection of narrative poems and one of our richest sources of information about Norse mythology, not to mention the ancient Germanic languages. The oldest version comes from Iceland and dates to the thirteenth century CE, though many of the poems are thought to be older. In the following verse the valkyrie Brunhild advises the dragon-slaying hero Sigurd on how to boost his chances of victory:

> Victory-runes you must inscribe
> if you want to have victory,
> and inscribe them on a sword's hilt,
> some on the battle-boards,
> some on the slaughter-cords,
> and name Týr twice.[2]

From the moment that these languages were written down, they became amenable to study. Linguists could chart their evolution and reconstruct their ancestors. But that moment, the dawn of history, is an example of what astronomers call an 'event horizon': it sucks in all the light. In many ways it's an optical illusion, because the wellsprings of European language, and culture, lie a long way behind it – in prehistory.

Hortobágy National Park is that fragment of the Hungarian *puszta* or steppe that the government deems worthy of protection, having turned the rest of it over to cultivation for its good, black earth. It feels different to the steppe east of the Don – the steppe where Natalia Shishlina excavates. The smells are different, as is the insect life. There are wetlands frequented by herons and occasional copses. In spring the birdsong is cacophonous. The puszta is a place where farmers have always felt as at home as herders.

In May 2023, a team of archaeologists led by Marianna Bálint, János Dani and Volker Heyd began excavating a burial mound in Hortobágy. First they found the remains of two dozen Magyars, the steppe horsemen who reached Hungary in the ninth century CE, bringing the Hungarian language. Then they discovered a possible Avar who, if his ethnicity was confirmed, came from the same direction three hundred years earlier. Beneath him lay two Scythians, and beneath them the individual for whom the kurgan was originally built. This young man had been placed in a typical Yamnaya death pose, on his back with his knees raised. Two flint arrowheads were mingled with his remains: one between his ribs and the other in the region of his pelvis. As they teased the soil away from the arrowheads, the archaeologists wondered if there might be one more burial beneath his: that of a farmer whose people had placed him in the ground before the first nomads arrived from the east. The Yamnaya often built their kurgans over older graves as a way of marking their territory. Those who came after them reused their kurgans for the same reason. This time, however, they found no farmer. The young man felled by arrows was the oldest burial in the sequence.

The kurgan that produced this rich layer cake of humanity lies to the east of the River Tisza, a tributary of the Danube that flows from north to south across Hungary. At about the same time that an intrepid party of Yamnaya struck out from the eastern end of their range, heading for China, others departed from the western end. The eastern and western groups probably spoke different dialects, and the westerners may have had some knowledge of farming that the easterners lacked, but they shared the same burial rite. They were both Yamnaya.

The migrants targeted a handful of destinations in the west. Many followed the Danube inland (as the farmers had, thousands of years earlier). Some arced away to the south, but the majority kept going towards the plunging gorges known as Iron Gates. Beyond these gorges and the mountains that border them, a cluster of Yamnaya kurgans erupts from the Serbian plains. The largest cluster of all, however, is further north, in Hungary. The first migrants were followed by others. For five hundred years they kept coming. Nobody knows how many inhabited the puszta at the migration's peak, but judging by the number of kurgans they left behind, János Dani says that forty thousand is a reasonable estimate.

Those migrants had travelled up to thirteen hundred kilometres (eight hundred miles) to get there – not as far as their eastern relatives but a very long way nevertheless. There is a debate over exactly who they were, with some archaeologists arguing that the first Yamnaya, at least, came in families, and others that roving male bands led the way. The role that horses played is controversial too. One way to reconcile the various strands of evidence is to think of the pioneers being predominantly male and mounted, with families following soon after, in wagons and on foot.³ The scouts may have been prospecting for metal. Irony of ironies, given the wonders once wrought at Varna, it was the Yamnaya who brought skilled bronze-working to the heart of Europe.

They may have been drawn in the first instance by the plentiful supply of timber in the forest-steppe, which they could use as fuel for smelting. The existence of the rich pastureland further west, inside the Carpathians, may have been a serendipitous discovery. Or perhaps they came expressly for the women and the cattle. Their burial rite remained the same, as time went on, but the buried did not. They became older and less male. Some of the women interred beneath the kurgans were tall and robust, displaying the typical Yamnaya build, but others were small and fine-boned. Besides the silver hair rings and dog-tooth pendants, the signature red ochre, the grave goods began to include exotic kinds of pottery. Some Yamnaya men seem to have found themselves local wives.

On the Danubian route the migrants stopped at the Tisza.⁴ Another stream headed north of the Carpathians, following the Dniester and passing through the so-called Moravian Gate to reach Bohemia in what is now the Czech Republic. Together, these two streams of Yamnaya constitute the last and largest of Gimbutas' waves, the one that most scholars agree brought the Indo-European languages to Europe. But the Yamnaya went no further. It was their descendants who carried their genes and their languages west to the Atlantic.

Around 2900 BCE – as the Yamnaya's descendants settled into the Altai and bands of Anatolian-speakers were learning to farm in Türkiye – a new culture arose in the lands between Bohemia and western Ukraine. Archaeologists call it Corded Ware because its practitioners decorated their pottery by pressing string into wet clay. The exact relationship between these people and the Yamnaya is unclear, but at least some Corded Ware people were the fruit of unions between the Yamnaya and the farmers they encountered in the forest-steppe.⁵ They resembled the Yamnaya in many respects, notably in their taste for milk and their impressive physiques. They buried their dead in a similar way too, with one striking difference: Corded Ware people placed stone battleaxes in their graves. Because of those axes, archaeologists generally consider them to have been more violent than their Yamnaya forebears. 'We are fated to beget a new and violent tribe . . .'⁶

The Corded Ware culture spread across the top of Europe as far as Belgium. This was an overwhelmingly male migration, one almost certainly spearheaded by warbands, and it moved, initially, at a staggering pace. Within fifty years the bearers of this warlike culture had literally blazed a path from Poland to Jutland. Burning forest as they went, they cleared expanses of heath on which they grazed their herds. 'They simply recreated the steppe in north-western Europe,' says Danish archaeologist Kristian Kristiansen. Before long they had reached the Rhine, and somewhere along that river's course they encountered a group of people who had travelled up from Iberia, probably in boats following the waterways.

These southerners are named for the highly decorated cups, shaped like upside-down bells, that they brought with them from Spain: Bell Beaker. The cups were so big that they had to be held in two hands, indicating that their purpose was ritual. Bell Beaker people buried their dead facing east, prompting some to claim that they worshipped the sun. They were skilled copper-workers and mariners, and very handy with a bow and arrow. For some archaeologists they were missionaries, for others tinkers, and it's not out of the question that they were both. Whoever they were, they almost certainly spoke a non-Indo-European language. Their ancestors had been in Europe far longer than the immigrants they now encountered on the banks of the Rhine.

The Bell Beaker and Corded Ware populations interbred and their descendants radiated out again, carrying steppe ancestry and the Bell Beaker culture into the furthest redoubts of Europe. Scholars think that these 'reborn' Beaker people spoke Indo-European languages, because they are the best candidates for carrying those languages through much of the continent. From the Rhine Valley, archaeologists track their material remains as far east as Hungary, south to Spain again, north to Norway and west to the Atlantic coast of France. Braving the English Channel in boats, they arrived in Britain around 2450 BCE and Ireland a few hundred years after that. One early Beaker immigrant to England, dubbed the Amesbury Archer, was buried amid great riches within walking distance of Stonehenge. The famous bluestones had not long been hauled into place, by previous inhabitants with different beliefs, but he may have stood in the centre of the ring at dawn on the winter solstice and experienced his own epiphany as the sun's rays flooded in.[7]

The Yamnaya culture vanished from the archaeological record around 2500 BCE. A hundred years later the climate began to cool again in Europe and the Corded Ware culture faded. The effects of this global climate event varied across the continent, but by 2200 BCE parts of Europe were experiencing severe drought and it was the Bell Beakers' turn to disappear (though their culture persisted for longer in Britain and Ireland). By 2000 BCE, as the drought began to ease, the great movements of people that had transformed Europe over the previous millennium had settled down. The dominant subsistence mode was farming and people were mostly sedentary, but for all that things looked similar on the surface, the social, economic and biological changes had been profound.

Genetically, the population now resembled that of modern Europeans, with its three-way mix of ice-age hunter-gatherer, Near Eastern farmer and steppe herder. Most men carried Y chromosomes that had originated in the steppe. People kept larger herds and wore woollen clothing. Lactose tolerance was spreading and meat represented a higher proportion of the diet. There is some evidence that dietary changes, or an incoming mutation from the east, or a combination of the two, led to a slackening of the lower jaw muscles. Overbite became more common, and with it sounds such as *f* and *v* that are made by pressing the lower lip to the upper teeth. The speech soundscape may have shifted subtly.

Bronze drove the economy and mining was a recognised profession. Huge volumes of tin, copper and finished bronze articles moved along the continent's trade routes, both terrestrial and maritime. Other goods were trafficked too, including humans, and the elite brotherhoods of old found new purpose in defending the precious cargo. Their alliances, forged in teenage rites of passage and reinforced by guest-friendship, **ghostis*, guaranteed them safe passage. From the Atlantic to the Caspian a new social model dominated: hierarchical, patriarchal, warlike. Europe had entered the Bronze Age, and scattered through it were sizeable groups of Indo-European-speakers. The daunting task before historical linguists is to bridge the gap between those preliterate peoples and their distant descendants who, many centuries later, scratched their names into stone. To do so they have to work backwards, from those first hesitant scrawls, to the birth of Italic, Celtic and Germanic – and beyond.

'Hark! Hark! The lark at heaven's gate sings.'[8]

When Germanic was first caught in the candle flame of writing, in the second century CE, there was only one runic script and one Germanic language. That language was spoken in a relatively compact area centred on the Jutland peninsula, extending southwards towards the Alps, and it had yet to fragment. Linguists consider that it was still very close to Proto-Germanic. They disagree as to when Proto-Germanic was born (most say 500 BCE, a few put it as early as 2000 BCE), but the prevailing view is that it developed out of dialects that arrived in the region with the first Corded Ware warbands.⁹

Italic and Celtic looked quite different at the moment they were first written down. The Italic languages had already diverged into Latin, Oscan, Umbrian and possibly Venetic – a language which, as its name suggests, was spoken in the north-eastern corner of the modern country, around Venice. Since no early Italic inscriptions have been found outside that country, the parent language, Proto-Italic, is thought to have been born in or close to it. Its date of birth is usually fixed, rather loosely, at sometime before 1000 BCE.

Celtic was also mature by the time it was etched into stone, but it was spoken across a much larger swathe of the continent. Early Celtic inscriptions record three distinct languages: Gaulish in Gaul, Lepontic in northern Italy, Celtiberian in present-day Spain and Portugal.¹⁰ Celtic is presumed to have been spoken further east as well, in part because the Greeks and Romans – Europe's first historians – said it was. Linguists consider that the common ancestor of all the Celtic languages was spoken at roughly the same time as Proto-Italic, but *where* it was spoken is a more difficult question to answer, because of the huge territory that Celtic-speakers filled when their languages hove into view.

Proto-Celtic has been pinned on the Atlantic seawall in the west, in the Austrian Alps in the east, and on the border of modern France and Germany in between. The last theory is the leading one today, in part because the French–German borderlands boast the highest density of Celtic place and river names. On the grounds that such names resist change because they serve a valuable function as signposts, some linguists consider them useful indicators of where ancient languages were spoken. The Rivers Main and Meuse were both named for Celtic deities, while the Neckar probably took its name from a Celtic root *nik*, meaning 'wild water'.

Another reason to place Proto-Celtic there, and not closer to the sea, is that its reconstructed vocabulary contains very few words related to maritime technology. Linguist David Stifter reports that Proto-Celtic had to borrow words for 'ship' and 'sail', suggesting that its speakers were landlubbers.

Germanic, Celtic and Italic are related by common descent. This is evident from their grammar, their pronunciation and their core vocabulary (English *father – mother – brother*; Old Irish *athir – máthir – bráthir*; Latin *pater – māter – frāter*). But the relationships between them aren't equal. Celtic and Italic are generally considered to be closer to each other than either is to Germanic, like twins with a third sibling. The first two form superlatives in the same way, while Germanic does it differently.[11] Germanic also has a whole class of verbs, the so-called modal verbs, that the other two lack. These are verbs that get placed before another verb, in its infinitive form, to express possibility, intent, ability or necessity (English examples are 'must', 'shall' and 'could'). Some linguists suspect that Italic and Celtic arose as a single, possibly short-lived language, Italo-Celtic, while Germanic arose separately. And they think that all three split from Proto-Indo-European early on, before the centum–satem (hard *k* to soft *s* before certain vowels) switch.

Loanwords help to place these prehistoric languages in time and space too. These fall into three categories: loans that Italic, Celtic and Germanic made to each other; loans that they received from sister branches that expired before they could be written down; and loans from the non-Indo-European languages that dominated the continent when the first Indo-European-speakers arrived.

Of the loans that the three surviving branches made to each other, the borrowings are more obvious between Germanic and Celtic than they are between Germanic and Italic, as if the first two had been closer in space. Relatively early on, for instance, Celtic donated its 'king' word, **rīg-*, to Germanic, where it became **rīk-* (the root of German *Reich* and Dutch *rijk*, meaning 'empire').

We know this word came from Celtic because, of the three branches, only Celtic converted the $ē$ in Proto-Indo-European *$h_3rēǵs$ to an $ī$ (the Italic branch kept the $ē$, as in Latin *rēx*). Think of Vercingeto*rix*, the Gaulish king or 'supreme king', to translate his name literally, who led an unsuccessful revolt against Julius Caesar in 52 BCE.

The dead Indo-European sister languages left their ghostly mark, curiously, in a clutch of words for animals with big feet: Dutch *pad* (toad), Irish *pata* (hare), Welsh *pathew* (dormouse).[12] These loans also hint at the early proximity of Germanic and Celtic, which must have been close enough to borrow from the same lost language. Then there are the loans that came from the Neolithic farmers of Europe. From these natives, most likely from the women they abducted or took as wives, the Indo-Europeans absorbed a host of words for plants and animals that were rare if not absent on the steppe. They included 'lark', 'blackbird', 'turnip' and possibly a word meaning 'bull' (**tauro-*, the root of Minotaur and toreador).[13] The vocabulary tracked the distribution of the flora and fauna it described. In the balmy south of Europe the immigrants took up words that would become *cupressus* (cyprus), *ficus* (fig), *lilium* (lily) and *rosa* (rose) in Latin. From the foragers who haunted the wind-blown dunes of the north Jutland coast they acquired a word that became Proto-Germanic **selhaz*, and eventually English 'seal'.

These loans, which are still in use today, are older than the Indo-European languages they enrich, since the farmers brought them to Europe thousands of years earlier, from the Near East (many may have been Hattian in origin). A few may be older still, if the farmers borrowed them in turn from the hunter-gatherers who inhabited Europe when they arrived. And along with the words came knowledge. Now, at long last, the Indo-Europeans acquired a word for 'bee' (**bhi-*), probably because their indigenous wives taught their children the art of sylvestrian beekeeping – how to lure a swarm to a hollow tree or other crevice and periodically harvest honey from it.

Bringing the linguistic clues together with the archaeological and genetic evidence, some linguists propose that the immediate ancestor of all three branches – Italic, Celtic and Germanic – was spoken during the mining boom that launched Europe's Bronze Age. On the modern Czech–German border, which might have roughly coincided with the frontier between the Corded Ware and Bell Beaker worlds, stand the Erzgebirge or 'Ore Mountains' – the only place in continental Europe where copper and tin occur together. From 2000 BCE, this region was home to a number of towns that doubled as important centres of metal production and ritual. They would have attracted people from both worlds, who brought a kaleidoscope of still mutually intelligible dialects. (A famous relic of those prehistoric towns is the Nebra sky disc, a bronze disc with a blue-green patina, inlaid with gold symbols, that depicts a solar boat sailing across the celestial ocean. A replica of it floats high above our heads on the International Space Station.) Around 1600 BCE, boom turned to bust in the Ore Mountains, and people left the region in search of better prospects.[14] This could have been the moment of schism between Germanic dialects, which stayed in the north, and Italo-Celtic ones, which moved away to the south-east.

Some of the émigrés settled on the Hungarian Plain, along the old trade routes connecting the Baltic to the Aegean (the roads along which amber and bronze were carted, and much else besides). There they built a series of large and well-defended settlements lining a corridor formed by the Rivers Tisza and Danube. Like their Bell Beaker ancestors, these 'Hungarians' worshipped the sun, but unlike the people of Nebra, who had buried their dead, they cremated theirs, packing the ashes into urns which they then buried in fields. This distinctive death rite now began to spread in two directions – north-west across Austria, and south-west towards Slovenia and Italy. If an Italo-Celtic language spread with it, the thinking goes, then where the path forked, the nascent Proto-Italic and Proto-Celtic languages parted ways. It's just one theory, but the enigmatic and long-extinct Venetic language lends some support to it. Around four hundred Venetic inscriptions are known, and some linguists conclude from these that Venetic was not Italic, but something older — a relic of the ephemeral Italo-Celtic tongue (meaning that if you heard it spoken, you'd have an inkling of what the languages of the Roman emperor Nero and his nemesis, the Celtic queen Boudica, sounded like when they were one). Some Venetic inscriptions were found in Austria and Slovenia, exactly where Italic and Celtic might have taken leave of each other.

At about the same time that the Hungarian settlements came to be, very similar towns developed in northern Italy. Archaeologists see enough similarities between the two — notably the 'Urnfield' cremation style — to convince them that they were connected by trade and human traffic. Steppe ancestry had already reached northern Italy by 2000 BCE (from where it diffused gradually southwards, towards the future city of Rome), but if the people who first brought that ancestry spoke an Indo-European language, it was probably one of the lost ones. Linguists suspect that the forerunner of Latin arrived later, with the migrants from Hungary. By 1600 BCE, those migrants were settling in the Po Plain, close to the modern city of Parma, and from that time on the population of the region grew. The people who frequented its thriving markets, who also carried steppe ancestry, might have bartered in Proto-Italic.

The markets thrived until 1200 BCE — the ominous date that sounded the death knell for so many Mediterranean civilisations, including the Hittites and Homer's Greeks — but then both the northern Italian and the Hungarian civilisations vanished from the archaeological record. They might have suffered from the wider economic downturn, or perhaps a new pulse of migration out of Hungary triggered a crisis in Italy. Tens of thousands of people fled the Po Valley, scattering with their pottery and dialects to other parts of the Italian peninsula. As they went, linguists think that Proto-Italic split into Latin, Oscan and Umbrian. All three languages lived long enough to be written down. Oscan graffiti on the walls at Pompeii guided its inhabitants towards mustering points in times of siege.

The second stream of migrants from Hungary headed north-west across Austria, plausibly carrying the dialects that would become Proto-Celtic and lending a 'king' word to the early Germanic-speakers into whose orbit they now strayed. From its Rhine cradle, Proto-Celtic then expanded, fragmenting as it went into Gaulish, Lepontic and Celtiberian in the west and undocumented sister languages in the east.

For every scenario I've sketched here, at least one alternative exists. As with all the early branchings of the Indo-European family, uncertainty reigns — even if that is less true than it was in Marija Gimbutas' day. But perhaps the greatest outstanding mystery regarding Celtic is when it reached Britain and Ireland — the only places where, besides Britanny in France, it is still spoken today.

After the Beaker people came to Britain around 2450 BCE, their DNA replaced around ninety per cent of the local gene pool, and all of the Y chromosomes. The turnover was similarly dramatic in Ireland when they crossed the Irish Sea around two hundred years later. Such a dramatic genetic rupture was almost certainly accompanied by a linguistic one, and linguists are pretty sure that the Beakers introduced an Indo-European language to those islands, but they don't think that language was Celtic. The dates don't add up. The Beakers were gone from Britain and Ireland by 1800 BCE (though their genetic legacy lived on), and Proto-Celtic was born no earlier than 1500 BCE. Somebody else brought Celtic in, at which point the language of the Beakers atrophied and died. One suggestion is that it was farmers from what is now France, who crossed the Channel in large numbers around 1200 BCE, but some linguists think that date is still too early. They suspect that a later group of immigrants brought Celtic in, whom geneticists have yet to detect.

How and when Celtic got to Ireland is in many ways more mysterious still. Whereas Britain received waves of immigration after the Beakers, that left traces in its gene pool, Ireland did not. The genetic make-up of its population has barely changed since the Bronze Age, meaning that modern Irish people trace almost all of their ancestry back to those four-thousand-year-old archers with their copper daggers, their big cups and their (possible) veneration of the sun. The dilemma in the Irish case is this: if foreigners didn't bring Celtic to the Emerald Isle, how did it get there?

Rowan McLaughlin, an archaeologist at Queen's University Belfast, thinks that Irish people might have gone to Britain and fetched it back. He has built a database of all archaeological traces of human occupation in Ireland from the Neolithic on, and used it to construct a timeline of population change. At 2500 BCE, he says, about a million people lived in Ireland, and that number remained relatively stable until 800 BCE. Around that date it fell drastically, but the decline was reversed three hundred years later. Since the genetic profile of the population remained the same, McLaughlin suspects that some disaster, perhaps related to a period of very wet weather that is known to have drenched north-western Europe around 800 BCE, triggered an Irish 'brain drain' to Britain. Three centuries later, the descendants of those émigrés returned to their roots speaking a new language: Celtic.

Others think that Celtic implanted itself in Ireland much later. The latest date of all is proposed by the Dutch linguist Peter Schrijver. He has argued that the Beaker language could have survived in Ireland right up until the first century CE, when a small group of British Celtic-speakers crossed the Irish Sea to escape the Roman invasion of Britain. Having enjoyed some measure of economic and social success in their new home, he suggests, they became role models there. Indigenous Irish people aspired to be like them, and the most aspirational of all learned their language. Over time, Celtic penetrated more and more domains of activity, from the political to the military, the mercantile to the religious, until eventually it found its way into ordinary people's homes and everyday life. But ordinary people would have spoken that Celtic language with a strong 'Beakerish' accent, and so what came out of their mouths was something new – the language we call (Old) Irish.

The linguistic landscape of Europe continued to reconfigure itself after the advent of writing, and we'll return to chart those transformations later. First, though, we turn to a more pressing question. Europe was densely populated with farmers when the first speakers of Indo-European arrived. So how did the outsiders' languages spread?

You have to marvel at it. When the Yamnaya arrived in Hungary, the population of Europe was an estimated seven million. The migrants might have numbered in the tens of thousands, yet within a thousand years languages descended from theirs were spoken across the continent.

To give a modern parallel, it's as if sometime in the early twentieth century, New Yorkers had stopped speaking English and started speaking Italian. As if all seven million of them, including the Puerto Ricans, the African-Americans, the White Anglo-Saxon Protestants, the Irish Catholics, the eastern European Jews and all the others, had begun cussing and haggling and praying in Italian. Italian immigrants started arriving in the United States in the late nineteenth century, and by 1930 – when New York City was the world's most populous city – they accounted for seventeen per cent of its population. That's a far higher proportion than the steppe herders represented in Bronze Age Europe, however vague our estimates. And of course New York did not flip. The Italians assimilated and learned to speak English, like everyone else.

This was the conundrum that Gimbutas and Renfrew came up against constantly in their public and private conversations, butting their heads against it, circling it, prodding it and eventually resolving it in two radically different ways. Renfrew pulled back the arrival of the Indo-European languages in Europe by three thousand years, to a time when the local population was less dense. He had them roll in on a tide of farming. Gimbutas envisaged a violent, male-dominated invasion. Most scholars, including Renfrew, now think that she was right in substance: the steppe people came and imposed their languages. But they think that she was wrong about the mechanism. So how did they do it?

In 2015, when the two studies were published that unequivocally demonstrated a massive genetic turnover in Europe around five thousand years ago, headlines in the mainstream press yelled that the Yamnaya roared west in an orgy of violence. Even then, many archaeologists felt uncomfortable about that conclusion. To Volker Heyd and his Hungarian colleagues, it was clear that the Yamnaya had not crossed the River Tisza. With few exceptions, they had not left the steppe to which they were so well adapted. They may have pushed further west on the route north of the Carpathians, but not much, and *pace* the young man with the two arrowheads in his flesh, they weren't spectacularly violent.[15] Traumatic injuries were relatively scarce in their graves, as were weapons. Their Corded Ware relatives, who streamed across the top of Europe wielding stone battleaxes, were probably more violent, but they hadn't achieved the linguistic transformation of Europe alone. The Beaker folk had played their part too, and they were only half-steppe by ancestry.

There was no getting round the genetic evidence: by 2000 BCE steppe ancestry had penetrated as far west as Ireland. In Britain and Scandinavia the genetic turnover was over ninety per cent, on the Iberian peninsula forty per cent, but in all those places the replacement of the male sex chromosome was near-total. By fair means or foul, the migrants had bred with local women and prevented local men from passing on their genes. Rape, murder, even genocide could not be ruled out, but as the geneticists gathered more samples and honed their methods, they began to see that other explanations could and should be entertained.

Most importantly, they became aware of the factor of time. If the replacement of the gene pool took sixteen generations rather than six, as it did in Britain after the arrival of the first Beaker people, it need not have been violent. You could envisage ripples of immigrants into a sparsely populated area, and the newcomers trading politely with the natives for decades or even centuries before the two began to intermarry. By then, the immigrants might not have seen themselves as immigrants any more, since generations of them would have been born there. Consider that a third-generation immigrant today rarely speaks their grandparent's native tongue and looks upon that grandparent's homeland as a holiday destination. Or that within two hundred years of the voyage of the good ship *Mayflower* to New England, its passengers were being fêted as Pilgrim Fathers, founders of a new nation. Their descendants were patriotic Americans.

One effective way of depopulating a landscape is disease, and since biologists can now read the genetic code of ancient microbes as well as that of the humans who hosted them, we know that the plague bacterium, *Yersinia pestis*, was already present in Europe at that time. (This was a major revelation of the last few years, since the sixth-century-CE Plague of Justinian was long thought to mark the beginning of the first global plague outbreak.) There has been disagreement over how deadly or transmissible the prehistoric form of plague was, compared to its medieval counterpart, but Kristian Kristiansen and the Copenhagen geneticists are now fairly sure that a dangerous epidemic, or epidemics, swept Europe in the late Neolithic. Plague wasn't the only human disease of animal origin circulating then either, and it's possible that the steppe migrants

– who lived in such close proximity to their herds – brought those diseases in. This was thousands of years before medicine moved out of the realm of magic; the death toll would have been high. Some estimates suggest that the Neolithic population of Europe was halved.

The Yamnaya brotherhood that ventured out of a Pontic river valley around 3300 BCE could have been the sole survivors of an early plague outbreak that cut down the rest of their community. If they survived because of a genetic predisposition that boosted their immunity, they would have passed that predisposition on to their descendants, some of whom would have entered Europe, along with the bacterium, and settled the not-long vacated pasturelands.[16] They may even have witnessed the ravages of the epidemic, interpreting them according to their own worldview. David Anthony has noted that both the Indic god Rudra and the Greek god Apollo hurled weapons that sowed disease, and both gods were associated with warbands. If that association already existed in the minds of the speakers of Proto-Indo-European, then – sobering thought – the advance guard might have imagined their gods clearing a path for them, especially if they themselves remained healthy.

The Yamnaya scouts need not have resorted to violence in that scenario. And there is another reason why they may have been able to expand their territories peacefully. A German anthropologist called Martin Trautmann has argued that their physical presence would have been intimidating enough.[17] Ancient historians recount that the Germani and Celts were on average six centimetres (just over two inches) taller than Roman centurions – enough, they implied, to give the barbarians a psychological advantage. Yamnaya men were on average *ten centimetres* taller than the male farmers they encountered. They would have had a noticeably heavier build and deeper voices. Their facial features – lantern jaw, prominent nose, deep-set eyes – can't have failed to impress their slenderer, lighter-boned interlocutors, and they may have heightened the effect by sporting tattoos on their foreheads. Traces of the dyes they used have been found on their skulls.

There was probably violence in places, but in general scholars now agree that it wasn't the main motor of Indo-European success. Much more important was the way that this new civilisation organised and perpetuated itself. The Yamnaya brought from the steppe a suite of institutions that they had honed over time to maintain social cohesion over large distances and long separations. They had expansion built into them, as archaeologists like to say. These institutions, which they bequeathed to their descendants, are reflected both in the physical traces they left behind and in the deepest layers of the Indo-European languages.

They began in childhood. From the age of about seven boys were fostered out. This was a way of defusing tensions potentially arising from the accumulation of sons in one household, but it also cemented alliances forged through marriage. Though a woman moved into her husband's home, her sons were sent to her kin, usually one of her brothers. A boy might receive military training from his maternal uncle-slash-foster father, to whom he was often closer than to his biological father. In Old Irish, the words for 'mother' and 'father' (*máthair, athair*) referred to one's real parents, while the more affectionate diminutives *muimme* and *aite* were reserved for one's foster parents. European mythology is brimming with foster brothers who were also best friends or even lovers, from Achilles and Patroclus in the Greek-speaking world to Cú Chulainn (pronounced Koo Hullin) and Fer Diad in the Celtic one. At about fourteen, following the age-set system, boys born to the elite headed into the wilderness to prove themselves, returning at twenty-one to take their place in society. But they maintained their links to each other all their lives long, forming alliances when the need arose and maintaining reciprocal ties of guest-friendship, sometimes over generations.

Geneticists have observed what could be the genetic signature of these customs. To begin with, the Corded Ware and Bell Beaker populations were extremely genetically diverse, as if young men from different clans had taken mates from outside the group, perhaps in the newly occupied territories. In both cases the genetic diversity eventually collapsed, and one or a few Y-chromosome lineages came to dominate in the male population. That could reflect the formation of an elite, as strict rules of marriage and inheritance concentrated power and wealth in the hands of a few clans. The ruling lineage was different in the Bell Beaker and Corded Ware worlds, as if a clan that had been lowly in Corded Ware had gone on to distinguish itself, earning elite status in Bell Beaker. It's possible that competition between the adolescent warbands for glory and loot shuffled the social hierarchy, or that a warrior elite in Corded Ware gave way to a priestly one in Bell Beaker. A similar realignment may have taken place earlier, on the Pontic steppe, producing an elite that we call the Yamnaya.[18]

A recent German study offers a glimpse of Indo-European life in the microcosm of one south German valley – a bug's eye view, if you like, of the customs and conventions that propelled the Indo-European languages through Europe. Around 2500 BCE, the floor of the Lech Valley was dotted with farmsteads, each of which had its own cemetery. From teeth found in these cemeteries, the researchers were able to build a portrait of the valley's inhabitants over the next eight hundred years – over the period, that is, when Europe was being genetically remodelled, and the Copper Age was giving way to the Bronze. In mesmerising detail, they described those people's kinship links, marriage practices and health, even their attitudes towards love and children.

Analyses of the isotopes in their tooth enamel revealed that some of the women buried in the cemeteries had grown up hundreds of kilometres away. If these 'foreign' women had had children, their children weren't buried with them. A second category of woman had grown up closer by, perhaps in the next valley, and with the exception of their adult daughters, their children *were* buried in the same cemeteries as them. Based on the fact that the study's authors found no half-siblings in the cemeteries, meaning that all siblings shared the same mother and father, they concluded that the two categories of women were probably co-wives (rather than wives who succeeded each other monogamously), and that these co-wives had fulfilled different roles. The foreign or 'social' wife might have brought prestige and connections, while the local one bore the children. The social wife wasn't necessarily childless; she might have left her own children in the place that she came from, or fostered them out. But once in her new home she was available to help raise those of her co-wife. Both wives seem to have been equally valued, since they were buried with grave goods as numerous and as costly as those of their husband.

Philipp Stockhammer, an archaeologist who worked on the Lech Valley study, pictures the male head of each farmstead travelling once a year to a town in the north, perhaps one in the thick of the Bohemian mining boom, and negotiating husbands for his marriageable daughters and foster homes for his younger sons. Apart from its social benefits, this would have been an effective way of moving knowledge and skills around at a time before literacy and technical schools. Stockhammer and his co-authors proposed, controversially at the time, that the foreign wives might have served as repositories of technological know-how. If the daughter of a Bohemian smelter, weaver or potter had observed her parent at work, taking an ever-greater hands-on role as she grew up, she would have been in a position to pass those skills on to her husband's family. None of those crafts require particular physical strength.

Few men would have dispatched their seven-year-old son or sixteen-year-old daughter on a five-hundred-kilometre journey alone, with wolves, bears and would-be assassins lurking behind every tree, so the fact that such journeys were routinely undertaken suggests that there was a network of roads, fords and inns in place. The human and bronze cargo probably travelled in the same caravans, chaperoned by warriors who recognised each other from their teenage brotherhoods or flashed their swords (whose hilts bore symbols of fraternity or other group identity) as passports. Armed protection notwithstanding, such a system cannot have operated for eight hundred years against a backdrop of constant conflict. Violence may have erupted from time to time, but peace was probably the default.

Polygamy – or rather polygyny, which refers to a man having many mates rather than many wives – was by no means an Indo-European invention. It was the norm in the ancient world, and even by the time of the Greeks and Romans, for whom marriage was supposed to be monogamous, societies condoned married men sleeping with other women and acknowledging the offspring of those unions.[19] It's possible, however, that the variety practised in the Lech Valley goes some way to explaining the success of the Indo-Europeans in imposing their languages and customs.

Life expectancy was much lower then, with the average woman unlikely to survive her thirties. But if a child's biological mother died, in a Lech Valley farmstead, there were other women available to step in as surrogates and ensure that child's survival. There was the presumed co-wife, but there was also a third category of woman found in the cemeteries. The women in this category had grown up locally but were not related to anyone else in the cemetery where they lay. They received simple burials with no grave goods. The study's authors suspect that they were domestic slaves.

Slavery might already have been common in Europe four thousand years ago, but it probably took a different form from slavery in, say, Roman times. One reason for thinking this is linguistic. The Latin word *famulus*, meaning 'slave', had a counterpart in Oscan *famel*, indicating that speakers of the Proto-Italic parent language already had a concept of 'slave' and a word to express it:

famelo-. But *famulus* is closely related to the Latin word for 'household', *familia*. That in turn suggests that a slave might have been regarded as an integral member of the household. She or he didn't enjoy the same status as the children of that household, but children might be asked to perform menial tasks too, and both were subject to the authority of the *pater familias*. Both might also have been sent out with the caravans, to worlds beyond their ken. The boundaries between free and unfree would become more sharply defined, in time, but in Bronze Age Europe they may still have been quite fluid.

If humans were travelling large distances, so were the germs that infect them. One in ten of the skeletons in the Lech Valley cemeteries was infected with the plague bacterium, indicating that the disease was endemic there. Similar finds across Europe suggest that plague outbreaks followed the waterways, those eternal routes of communication and trade. Yet the steady influx of people from distant places, in the form of wives and foster sons, could also be the reason that the community survived the outbreaks. By replacing the dead they boosted clans' resilience. They also, potentially, boosted their fertility. If, as Stockhammer thinks possible, the social wife sometimes breastfed children who weren't hers, she might have 'liberated' her co-wife to conceive again sooner, since lactation can act as a contraceptive.[20] The Indo-Europeans may simply have been good at having children and keeping them alive. If they kept it up over generations, steppe ancestry would have spread through the population, and Indo-European languages with it.

Studies of kinship and custom like the one in the Lech Valley are transforming prehistorians' understanding of how people moved through the ancient world. Epic treks led by men on horseback may have contributed to the turnover in the European gene pool and to the spread of languages, but a more important factor was probably the cumulative, smaller displacements of women and boys – the human trafficking on a no less epic scale that defined Indo-European-speaking societies from an early date. And though some men may have fathered many children, the impressive genetic legacies of a handful of individuals – including the patriarch of the first Yamnaya clan – probably owe more to the naturally exponential growth of their lineages over generations than to their personal sexual prowess.[21]

It is not possible to conclude a discussion of what allowed Indo-European languages to spread without dwelling awhile on booze. Even before the Neolithic, humans had almost certainly experimented with natural sources of intoxication such as rotting fruit, and been successively spooked and elated by their effects. It has even been argued that religion was born of those experiments: prehistoric people walked through the doors of perception and found gods on the other side. Farmers with roots in the Near East are thought to have been the first to control the fermentation process, but it was people from the steppe who turned drinking into a social obligation.

Alcohol is, unfortunately, difficult to detect in the archaeological record. One day soon it will be possible to identify the proteins of bacteria used in fermentation, but until now archaeologists have mostly looked for drinking cups that contain traces of grain, grape or the pollen that gets trapped in honey. This line of investigation, combined with linguistic and mythological references to drinking, persuades them that, at least in Europe, alcohol was at the heart of Indo-European rituals. It was the lubricant that eased the passage of the Indo-European languages through an expanding web of alliances.

Historical linguists combing through the strands of the Indo-European diaspora surmise that it might all have started with a honey-based beverage, since words derived from *medhu* (mead) are ubiquitous. In Odin's hall, Valhalla, the valkyries served horns of mead to fallen warriors, and when Zeus wanted to make his father vomit up his siblings, he gave the crazy old Titan a laced honey drink. Fermented mare's milk may have been another early tipple. Wine and beer joined the drinks cabinet in time, and were assigned their own god (Dionysus to the Greeks, Bacchus to the Romans). But the intoxicating liquid supped throughout the *Rig Veda* was probably not alcohol. The nature of *soma* has eluded scholars for centuries. It clearly had hallucinogenic properties, causing some to associate it with a mushroom, but since it also induced euphoria others favour a plant with amphetamine-like properties, something like the *khat* or *coca* leaves that people seeking a high still chew today. 'I am huge, huge! flying to the cloud,' boasts a god or sage in the *Rig Veda*. 'Have I not drunk soma?'

Drinking was not decadent in the Indo-European world, or not to begin with. It was symbolic, inseparable from the most important rites and contracts. It sealed oaths and consecrated kings. 'He will not be a king over Ireland, unless the beer of Cuala comes to him,' states the set of Irish myths known as the *Kings' Cycle*. The nectar of the gods conferred immortality, but also wisdom and poetic inspiration, which is why, in Indo-European mythology, it was always being stolen. The Russian firebird (whose capture requires the hero to undertake a daunting quest) and the Greek myth of Prometheus (who took fire from the gods and gave it to humans) are two riffs on the Indo-European theme of stealing the fiery juice from heaven. *Promētheús* originally meant 'the one who steals', not 'forethought' as it is sometimes translated. Its exact counterpart appears in a Vedic myth about the theft of fire: *pra math-* ('to steal').

Whether the Yamnaya drank alcohol is not clear. They may have preferred to smoke cannabis, since they scattered its charred seeds all the way into Europe (the farmers they encountered seemed to prefer the narcotic effects of the opium poppy). Drinking culture looks to have spread through Europe with the Corded Ware and Bell Beaker folk, and it was perpetuated by their heirs, the Celts, Romans and Germani, as well as the Balts and Slavs. At some point, probably as early as the third millennium BCE, drinking alcohol became associated with the two-handled tankard. Once this object had made landfall in Europe, it never went away. When the English yelled *Læt hit cuman!* ('Let it come!') as they caroused on the eve of the Battle of Hastings, they weren't referring to William the Bastard's army but to the loving cup, the two-handled tankard that is still passed around after dinner at certain Oxbridge colleges.

It was principally *social* mechanisms, in other words, that drove the spread of the Indo-European languages. The seeds were sown in raids, marriage, fostering and patronage. They were tamped down through storytelling and watered at feasts. Seeds became trees which became plantations which became islands. If there was undoubtedly some coercion at first, in any given region a tipping point would have been reached when so many people spoke an Indo-European language that those who didn't switched voluntarily. First they became bilingual, then they stopped speaking the old languages to their children. Parents calculated that it would be better if the next generation was fluent in the language of power and wealth, or teenagers appraising the world decided this for themselves. The islands expanded and eventually joined up, producing the mosaic that existed when history began.

The stereotypical Celt wore a checked cape and breeches, bleached his hair with lime and spoke in riddles. The Greeks called them Keltoi, the Romans Celtae or Gallatae, but those early chroniclers were sometimes rather hazy about the difference between them and the Germanic tribes. For the Greek geographer Strabo, writing in the time of Caesar, Germani were just extreme Celts, taller, wilder and with even yellower hair. The Celts were 'war-mad' but not malevolent, and very fond of feasting. The Romans described them running into battle naked, wearing only the sacred torque or neck-ring that conferred on them divine protection, or so they thought.

In reality the Celts were not one people but a loose federation of tribes who spoke different Celtic languages. Their art was highly sophisticated, as was their metalwork, and they were skilled warriors. Though we don't know when Celtic languages reached Britain or Ireland, the Celtic language space probably reached its maximum extent around the third century BCE. In the previous century, Gauls had crossed the Alps into northern Italy, where their language had absorbed its Lepontic relative (citing earlier sources, the Roman historian Justin suggested that the pressure for the Gauls to expand was caused by overpopulation; it might have been spearheaded by a sacred spring). In the east, profiting from the confusion that followed the death of Alexander of Macedon in 323 BCE, Celts had moved out of the Carpathian Basin towards the Balkans, and in the next century they invaded Greece. One contingent, speaking a language related to Gaulish, crossed the Bosporus and settled in the highlands of central Anatolia, close to the old Hittite capital Hattusha.

Meanwhile, Rome was expanding its frontiers. When the city was founded, in the eighth century BCE, Italy was ruled by the Etruscans. The two peoples shared steppe ancestry, probably because Italic-speakers had married into Etruscan society, but the Etruscans spoke a non-Indo-European language. By 500 BCE the Latin-speaking Romans had thrown off the Etruscan yoke, and though the Etruscans continued to exert a strong influence over them, it was the Roman language, and Roman identity, that came to dominate their fused societies.[22] (The proof is that we know the Etruscans by the name the Romans gave them, rather than the name they gave themselves: Rasenna.)

For a time Oscan, the language of the Sabines, was spoken more widely than Latin, but by 250 BCE the Romans had conquered most of mainland Italy and Latin had smothered its sisters. In the next century they took northern Italy from the Celts, and after that they turned their attention to Greece. The Romans were in such awe of Greek learning, however, that the Greek language was spared the fate of Latin's other victims. In the last decades before the Common Era, the Roman poet Horace would write: *Graecia capta ferum victorem cepit* . . . (Captive Greece captured her savage conqueror . . .).

Greek was another Indo-European language (one whose idiosyncratic story we'll tell in Chapter Eight), but in absorbing hundreds of Greek words Latin moulded them to its own grammatical rules.[23] It then funnelled them into the territories it conquered, along with its own, as the Roman Empire grew. The Latin-speaking world expanded largely at the expense of the Celtic-speaking one, but it wasn't as simple as the defeated Celts meekly switching to Latin. The Celts of Iberia heard the approaching tread of Roman centurions from the end of the third century BCE, but Celtiberian is documented long after Iberia fell – right into the reign of the Emperor Augustus several hundred years later. The Celts of Anatolia were eventually absorbed into the Eastern Roman Empire and learned to speak its lingua franca, Greek, but for a long time Celtic remained their mother tongue. (Christ's apostle Paul would address them – in Greek – in his letter to the Galatians, and they live on in the Turkish football team Galatasaray, whose name means 'Palace of the Celts'.)

Gaulish, for its part, survived the Gallic Wars of the fifties BCE by at least four hundred years. Linguist David Stifter has even argued that it wasn't Caesar who signed its death warrant, but the Germanic tribes who swept in as the Western Roman Empire collapsed. In fact Vercingetorix may have had his posthumous revenge. Though the conquered cities of Gaul embraced Latin as their language of administration, Gaulish survived in the countryside, and it left its imprint in French. The French way of writing 'eighty' as 'four-twenties' (*quatre-vingts*), is a relic of Gaulish vigesimal counting in the otherwise decimal Roman system. As the Celticist Henri Hubert once put it, 'French is Latin pronounced by Celts and put at the service of the Celtic mind.' Nor did the

British convert to Latin after the Romans invaded in 43 CE. A bilingual elite emerged, through which hundreds of Latin words entered (British) Celtic. But the masses carried on speaking Celtic.[24]

Proto-Germanic stayed more or less quietly in its heartland until about the third century CE, when it split three ways – into northern, western and eastern branches. The northern group comprises all the modern descendants of Old Norse, the language of the Vikings and of the *Poetic Edda*: Norwegian, Swedish, Danish, Elfdalian, Icelandic and Faroese. Their western relatives include German, Dutch, Frisian and English, while the only member of the eastern group for which we have textual evidence is extinct Gothic. Gothic was probably the first to leave the Proto-Germanic homeland, from where it gravitated first to Romania and the lands north of the Black Sea, before spreading back west towards Gaul, Spain and Italy in its Ostrogothic and Visigothic dialects. Long after it had faded in the west, a variant of Gothic was still being spoken in its older stronghold of Crimea, where it died in the eighteenth century.

West of the River Elbe, Latin held on to its conquered territories even after its imperial scaffolding had crumbled. Not the arrival of Christianity, nor the Germanic invasions, nor a mini ice age, nor the Plague of Justinian succeeded in dislodging it. But it did change, first under the influence of the Celts and later under that of the Germani. The Germanic kingdoms that arose from the empire's ashes – besides the Ostrogoths and Visigoths, the East Germanic-speaking Burgundians and Vandals and the West Germanic-speaking Franks and Lombards – adopted the language of the masses, Latin. It was their barbaric pronunciation that drove Vulgar Latin into the various dialects that would flower as the Romance languages, Dante's *oc*, *oïl* and *sì*. 'Langue d'oc, ancient language of the south,' wrote Swiss poet Charles Ferdinand Ramuz in his paean to the River Rhône, 'you remain loyal to this waterway, and a rosary of dialects extends and spreads along these currents like beads made of the same wood . . .'[25]

Something different seems to have happened along the old imperial frontiers. Those Latin-speakers who lived close to the Rhine took up German. Towards Rotterdam, at the mouth of the river, the result was Dutch. Closer to the Alps, where the river rises, it was High German (as in 'Highland German'), now distinguished from the Low German spoken further north principally by the way consonants were pronounced. A Low German *p* became High German *pf* or *f* (giving *apfel* instead of *appel*, for 'apple'), *t* became *s* or *ts* (turning *twee*, 'two', into *zwei*), and *k* became *ch* (converting *maken*, 'to make', into *machen*).

East of the Elbe, Slavs followed the Goths south-east and their languages ousted Gothic. Only ten centuries later would Germanic reclaim some of the territory that it had lost in the east. The form that expanded then was High German, which benefited from being the language into which Martin Luther had translated the Bible, and the first language of the printed book in the German-speaking world.

In fact Germanic's only significant territorial gain after the fall of Rome was England. In the fifth century, Low German-speakers from the Jutland area headed along the North Sea coast and across the English Channel in large numbers.[26] Some British Celts fled north and west to escape them, implanting Welsh and Cornish where they found refuge, but the majority took up the immigrants' language and, speaking German with a Celtic lilt, invented Old English. (The divergence of Welsh and Cornish could have been accelerated by the exotic pronunciation of the Latin-speaking elite who followed hard on the Celts' heels, being also persona non grata in the new Anglo-Saxon power centres.) Hints that the Britons switched under duress are provided by the Venerable Bede, an English monk who described what happened a few centuries later. Angles, Saxons and Jutes, Bede wrote, 'began to increase so much, that they became terrible to the natives'.

Old English itself corroborates Bede and those few other monks who hinted at the violence unleashed on the British. On the continent, the Germani had referred to all foreigners, meaning Celts and Romans, as *walhaz*. *walhaz* is the root of 'Walloon', the term still applied to Europe's most northerly French-speakers, but also of 'walnut' (originally a nut from France). In Old English the word became *wealh*, which is the root of 'Welsh'. It still meant 'foreigner', but now it specified a Celtic-speaking Briton. And it acquired a chilling second meaning, of 'slave'. English preserved hardly any Celtic words, but among the few that it retained are *ass*, *hog* and *brock*, as in 'badger'.

In the eighth century CE, another wave of Germanic-speaking migrants started to arrive in England: the Vikings. But the Vikings, though they donated some words to English and tweaked its grammar, left a relatively light linguistic imprint. The reason may be that Old Norse and Old English were still similar enough that their respective speakers could communicate, albeit at a basic level, without bothering to learn each other's language (in the same way that monoglot speakers of Italian can get by in Spain today, and vice versa). Old Norse had vanished from England within a few hundred years, leaving a lick of salt in the English lexicon and some ineradicable place names – Grimsby, Skegness and Whitby among them. (The Vikings easily imposed their language on Iceland, on the other hand, because when they arrived, apart from a few Irish monks, there was nobody there.)

The next language to invade Britain, in the eleventh century CE, was another Romance language: Norman French. The Normans were only half a dozen generations removed from their Viking roots and relatively recent converts to French themselves (hence 'Norman', a corruption of *Norðmenn* or 'north men'). They still spoke it with a Nordic accent, and it was their form of French that became the new language of prestige in England, though never the language of the English. A pithy illustration of this is the debate that opens Walter Scott's 1819 novel *Ivanhoe*, in which a swineherd and a jester discuss how an *ox*, a *calf* and a *pig* are German in the farmyard and French once presented, apple in mouth, at the lord's table (beef from *bœuf*, veal from *veau*, pork from *porc*).

Linguists have argued over how Germanic English remained in its fibre, in the face of this onslaught from Romance, but the consensus is that it resisted. Though it received a huge injection of Latin (and Greek) words via Norman French, along with new rules about spelling and pronunciation, it remained recognisably English. George Orwell embraced an old canard (French for 'duck') when he lamented what he saw as the unwarranted substitution of good, honest Anglo-Saxon words by Latin and Greek fripperies, but as his own writings attest, they all legitimately belong to English. In the following excerpt from his novel *1984*, the words in **bold** are of Latin origin, those in *italics* are French, and the rest are Anglo-Saxon:

> 'What can you do, thought Winston, against the **lunatic** who is more **intelligent** than yourself; who gives your **argument**s a fair hearing and *simply* **persist**s in his **luna***cy*?'[27]

Not even Arthur, Celtic slayer of Saxons, escaped the French polish. Having been a mere lad of fighting age, a *cniht* in Old English, he acquired a suit of armour, a mount and chivalry. He became a *knight*. Arthur may only have existed in legend, or semi-legend, but the Celtic languages of Britain and Ireland did claw back some territory from encroaching Germanic. At around the time of the Anglo-Saxon invasions, Irish slavers established colonies in Scotland and the Isle of Man, implanting Scots Gaelic and Manx.[28] Cornish migrants crossed the Channel to establish a colony at the northwestern tip of France. There they may have encountered the last speakers of Gaulish, the language in which Vercingetorix had uttered his last words before Caesar had him put to death in Rome in 46 BCE. Under Gaulish influence, Cornish became Breton.

Celtic continued to thrive in its Irish stronghold, for a time. In the twelfth century, land-hungry Anglo-Normans crossed the Irish

Sea from England. They soon took up Irish, but the lower-class English traders and craftspeople who came in their retinues continued to speak their mother tongue. English declined in Ireland after that, but it was zealously re-seeded, along with Protestantism, with the 'plantations' of the sixteenth and seventeenth centuries – when the English Crown confiscated Irish-owned land and encouraged English people to settle on it. After the devastating potato famine of the 1840s, for which the colonial power was largely responsible, English made further, dramatic inroads on the island, and by 1900 Irish was a minority language there. Even Daniel O'Connell, the nationalist leader who had fought for Catholic emancipation, talked of Irish as having had its day. 'Although the language is associated with many recollections that twine round the hearts of Irishmen,' he rued, 'yet the superior utility of the English tongue, as the medium of all modern communication is so great, that I can witness, without a sigh, the gradual disuse of Irish.' English, which had started out as a language of the oppressed, had become an oppressor.

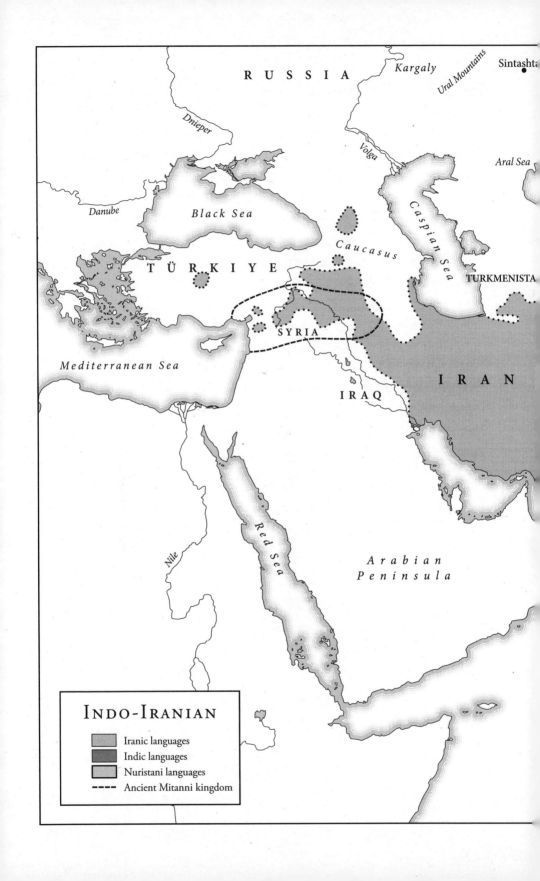

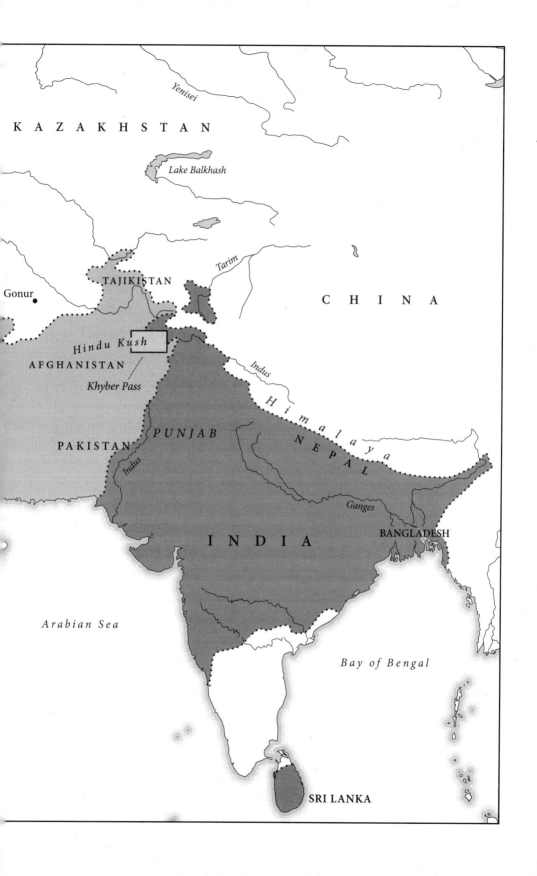

6

THE WANDERING HORSE

Indo-Iranian

In the spring, a stallion was chosen from among the king's horses and released to wander for a year. Four hundred of the king's warriors followed it, to ensure that it wasn't harmed or diverted. All the lands that it crossed during the year fell into the possession of the king, because the horse represented the king; it symbolised divine power. When at last it returned it was hitched to the king's chariot. Other animals were sacrificed and finally the horse itself was put to death. At the moment that the priest smothered it, he whispered to it: 'You do not really die through this, nor are you harmed. You go to the gods on paths pleasant to go on.' The queen then lay with the dead stallion beneath some covers; she simulated sex with it as the king's other wives stood around bantering obscenely with the priests. Finally the horse was dismembered, its meat divided into portions for gods and portions for mortals. 'Keep the limbs undamaged and place them in the proper pattern,' the *Rig Veda* instructs. 'Cut them apart, calling out piece by piece.' Thus concluded the most spectacular of the ancient Indian ceremonies, the *ashvamedha* or horse sacrifice.

In the late 1960s, when archaeologist Vladimir Gening and his student Gennady Zdanovich began excavating a four-thousand-year-old cemetery in the undulating steppes south-east of the Ural Mountains, they discovered the remains of horses along with some of the world's oldest chariots and hoards of weapons. Cattle, sheep and goats had been sacrificed too, but the sheer number of horses was striking: up to eight in a single grave. They had all been dismembered, their lower leg bones carefully separated at the joints. As incongruous as it seemed, in the middle of the Russian steppe, the archaeologists wondered if they were looking at the origins of the *ashvamedha*.

In time, they excavated twenty settlements belonging to the Bronze Age culture that they dubbed, after that first site, Sintashta. The cemeteries evoked ancient Indian ritual in other ways too. The grave pits had walls shored up by planks and wooden roofs held in place by timber posts. 'Let the earth as she opens up stay firm, for a thousand pillars must be set up,' says the *Rig Veda*, and: 'let me not injure you as I lay down this clod of earth'. In one sacrificial pit an upside-down pot had been placed between two rows of animal bones. 'Varuna has poured out the cask, turning its mouth downward,' the *Rig Veda* says of a god who, by this gesture, unleashes the rain. And then there were the burial mounds. 'House of clay' is a metaphor for death throughout the *Rig Veda*: 'Let me not go to the house of clay, O King Varuna, not yet.'

The existence of the Sintashta culture only came to international attention in the 1990s, when the Russian descriptions of the sites began to be translated into other languages, but they immediately caused a sensation. Abroad, too, scholars debated whether there could be links with the Indo-Iranian branch of the Indo-European language family. This branch, the largest in terms of both number of speakers and geographical range, comprises the Indic and Iranic languages as well as the Nuristani languages of the Hindu Kush. Today they are spoken by close to a billion people across South and South West Asia.

The Indo-Iranian languages are very different from Tocharian, the other eastern branch of the family. Most linguists agree that as many as a thousand years separated the departure of Tocharian and Indo-Iranian from the homeland, and that the *Arya*, as the Indo-Iranians referred to themselves, undertook a very different odyssey. But there the agreement ends. Some claim that farmers carried the Indo-Iranian languages across the Iranian Plateau, south of the Caspian Sea, more than five thousand years ago – as they rippled out from the epicentre of the Neolithic Revolution. Others say that the languages took a northerly route via the steppe much later. Sintashta could potentially decide the issue. If the people of that culture spoke a forerunner of Persian and Sanskrit, if they performed rituals that their descendants carried to India and Iran, then the Aryans must have crossed the steppe. Thirty years on, linguists and geneticists have added their contributions to those of the archaeologists. But the Aryan chapter of the Indo-European story is arguably the most politicised and contentious in the twenty-first century, and while some are satisfied that a solution is at hand, others say that nothing has been settled at all.

Persepolis was the ceremonial capital founded by Darius the Great in the sixth century BCE, and because of its remote location in Iran's Zagros Mountains it was known as the safest place in the Persian Empire to store art, archives and gold. Alexander of Macedon scuppered that reputation in 330 BCE, when he sacked and plundered the city and then set fire to it, destroying its many treasures and leaving it in ruins. Historians have long discussed whether or not the fire was deliberate. Alexander, Iskandar to the Persians – and 'accursed' rather than 'great' – might have wanted revenge for the Persian campaign against Greece a hundred and fifty years earlier. But he had nothing to gain from destroying the city and indeed, given both his imperial ambitions and his affection for Persian culture, much to lose.

One of the best-known accounts of the fire was written by the Greek historian Diodorus of Sicily a few centuries after the embers had cooled. Diodorus was in no doubt that it was vandalism. Once the city was theirs, he wrote, the invaders threw a party. Alexander made lavish sacrifices to the gods, the drinking went on late into the night, and 'as they began to be drunken, a madness took possession of the minds of the intoxicated guests'. An Athenian woman, Thaïs, suggested that they set fire to the city, to avenge the Persian destruction of the Acropolis, and others approved her plan. Diodorus went on: 'When the king had caught fire at their words, all leaped up from their couches and passed the word along to form a victory procession in honor of the god Dionysus. Promptly many torches were gathered. Female musicians were present at the banquet, so the king led them all out to the sound of voices and flutes and pipes, Thaïs the courtesan leading the whole performance. She was the first, after the king, to hurl her blazing torch into the palace.'[1]

The resulting conflagration spared only a husk of Persepolis, which gradually sank into the sand. The city was forgotten, erased from history. Within a few hundred years, however, travellers to the remote ruins had noticed the inscriptions carved into the columns that were still standing. Others would follow in their footsteps and make a more systematic study of the carvings. By the early nineteenth century they had unlocked the secrets of the script, cuneiform, and discovered that the inscriptions were written in three languages: Elamite (a language isolate, meaning that it belonged to no known language family), Babylonian (a dialect of Akkadian) and Old Persian, the ancestor of Farsi or Modern Persian. These were three of the official languages of the Persian Empire in the time of Darius, Persian being his mother tongue.

Around the time that the inscriptions at Persepolis were being decoded, another effort of decipherment was underway in India.

Having declared Buddhism the state religion in the third century BCE, the king Ashoka had peppered the subcontinent with billboards broadcasting his edicts. (In one of them he described, in the third person, how a particularly bloody war had led to his own conversion: 'From that time pity and compassion overcame him. And he bore it grievously.') These polished sandstone pillars were inscribed in colloquial forms of Sanskrit using the Brahmi script. This was the same script that Tocharian monks would use centuries later to copy sacred texts. Brahmi was deciphered in 1837, and from then on scholars could read and translate Sanskrit, the language of the Vedas. In India as in Persia, it was as if they had been handed the keys to the kingdom. They could begin, at last, to reconstruct the ancient histories of these great civilisations.

The oldest Indic text is the *Rig Veda*, the first of the four Vedas that make up the most ancient Indian, and Hindu, scriptures. Its thousand or more hymns were passed down orally for a very long time before they were recorded, and they were not recorded all at once. The dates of their compilation have been estimated based on clues within them. Place names suggest that they were written in the foothills of the Hindu Kush, in the Punjab region that straddles modern India and Pakistan. This was where the people of the Indus Valley Civilisation, the mysterious Harappans, built their cities nearly five thousand years ago, in another of humanity's early experiments in urbanism.[2] Those cities were abandoned around 1800 BCE, and since the Vedic poets knew them only as ruins, they are generally thought to have been at work after that. On the other hand, the poets did not know iron, which made its entrance into South Asia around 1000 BCE, though they did know copper and bronze. Scholars therefore hazard that the *Rig Veda* was written down between 1500 and 1200 BCE, at least four hundred years after the earliest Hittite texts.

But the *Rig Veda*, though the earliest literary text in Indic, is not the earliest recorded use of that language. That is a set of inscriptions dated to 1500 BCE and found in an unexpected place:

Syria, at the eastern edge of the Mediterranean – a very long way from the Punjab. At that time, northern Syria was ruled by the kings of Mitanni. Day to day, the people of the Mitanni kingdom spoke Hurrian, a non-Indo-European language, but for certain purposes they switched to one very like Sanskrit. Their kings invoked the gods of the Vedas to witness their peace treaties, and they took Sanskrit throne names. The name of the Mitanni king Tushratta is a corruption of Sanskrit *Tvesa-ratha*, meaning 'having a fearsome chariot'.

The Mitanni were expert breeders and trainers of horses, and when they were discussing this subject, too, they switched to Sanskrit. In the thirteen hundreds BCE, a Mitanni *assussani* or master horse-trainer named Kikkuli wrote a plan for rearing and training horses that is still quoted today because of the superb horses it turned out.[3] (An extract from his recipe for equine strength, translated by the Czech linguist Bedřich Hrozný, reads: 'Pace two leagues, run twenty furlongs out and thirty furlongs home. Put rugs on. After sweating, give one pail of salted water and one pail of malt-water. Take to river and wash down. Swim horses.') Though Kikkuli's name was of Hurrian origin, his manual was peppered with Sanskrit or Sanskrit-like words, including ones for the colours of horses: *bapru* (from Sanskrit *babhru*, 'chestnut'), *parita* (*palita*, 'grey') and *pinkara* (*pingala*, 'reddish').

For a thousand years Sanskrit dominated India and the parts of South East Asia that it reached through maritime trade. Later it spread even further afield, carried with Buddhism to the Tibetan Plateau, the Tarim Basin and China. On the subcontinent it was both the vernacular and the language of the custodians of Hinduism, the Brahmins. Abroad it spread as a literary language, and not just in the religious domain. The *Mahabharata* and the *Ramayana* are two epic Sanskrit poems concerned with princely trials and tribulations. There are Sanskrit works on philosophy, science, cookery, elephants, yoga and sex, in the form of the *Kamasutra*. 'Language, auspicious, charming, like a creeper, whose minds does it not win over?' as a *sukta* or hymn has it. Sanskrit probably died as a vernacular sometime in the first millennium CE, having disintegrated, like Latin, into a rich panoply of daughter languages. There are about a dozen of them in all, including those that are most widely spoken in South Asia today: Bengali, Marathi, Urdu and Hindi. But the association of Sanskrit with the sacred ensures that it continues to be used in religious contexts.

The oldest surviving text in any Iranic language is the *Avesta*, the holy book of Zoroastrianism. Zoroastrianism was the religion of Persia before the Muslim conquest of the seventh century CE, and as the first truly monotheistic creed it had a profound influence on the Abrahamic religions. The *Avesta* also came together over time, and its oldest hymns, the *Gathas*, are said to have been composed by Zarathustra himself (or Zoroaster to the Greeks). The places he named in them indicate that he lived in what is now eastern Iran, not far from Afghanistan. The language he spoke is known as Avestan. Closely related to Old Persian, Avestan may already have been restricted to the liturgical domain by the time Darius came to power.

Zarathustra was a reformer, and the *Gathas* can be seen partly as a critique of the *Rig Veda*. Though their age is also debated, because it's not certain when Zarathustra lived, it's generally thought that he composed them around 1000 BCE – a few centuries after the *Rig Veda* was compiled. He and the Vedic poets were very much aligned in their manner of thought and expression. Though he criticised their celebration of conquest and sacrifice, he referred to the same gods and rituals as them, and used the same poetic devices. Cows were constantly being stolen and sacrificed in the *Rig Veda*, and in one *Gatha* Zarathustra takes the cow's point of view, lamenting her rough treatment at the hands of raiders: 'Cruelty, oppression, bloodlust, rage, and violence have fettered me.'[4] He and they are so much on the same wavelength that many linguists think Sanskrit and Avestan might still have been mutually intelligible at the time that they were working, or at least that they had not split very long before. That split is usually dated to 2000 BCE at the earliest – roughly the date at which the Sintashta were sacrificing horses on the steppe.

Less is known about the earliest documented speakers of Iranic than Indic, because the oldest Iranic texts to have been preserved are exclusively religious. There is no equivalent of the *Kamasutra* or the *Mahabharata* to illuminate other areas of their lives. The Iranic languages are spoken today in roughly the region where they are thought to have arisen – Iran, Afghanistan and Tajikistan – but they expanded enormously before contracting again. At one point they were spoken from the steppe north of the Black Sea, southwest to Mesopotamia and east to the Tarim Basin. Because they filled that enormous space they were highly diverse, more different from each other than are the many daughters of Sanskrit. By the first millennium BCE they were spoken by everyone from the horseback warriors the Greeks called Scythians and their eastern relatives the Saka, to Sogdian traders and Persian emperors. The Persians left the world one of its great poetic traditions, spanning the love poetry of Hafez and Rumi, Ferdowsi's epic *Shahnameh* (*The Book of Kings*) and the quatrains of Omar Khayyam:

> The moving finger writes; and, having writ,
> Moves on: nor all your piety nor wit
> Shall lure it back to cancel half a line,
> Nor all your tears wash out a word of it.[5]

But the wide-ranging nomads of the steppe left more of a mark in the landscape. If you were struck, in Chapter One, by the repetition of the initial D in the names of the rivers feeding the Black Sea, know that it isn't a coincidence. An Iranic word for 'river' was *danu*, which is the root of both Don and Danube.

Dniester comes from *Danu nazdya*, 'river to the front', and Dnieper from *Danu apara*, 'river to the rear'. These names were the legacy of the Scythians who inhabited those parts in the Iron Age (leaving a burial mound in Alexey Nikitin's natal village) and who spoke languages not unlike the Pashto of modern Afghanistan.[6] The extraordinary Scythians travelled with their cats in felt tents mounted on wheels, accompanied by herds of horses. They were succeeded on the steppe by the Sarmatians and Alans, who also spoke Iranic and who were dispersed or absorbed by other groups when the non-Indo-European-speaking Huns swept through in the fourth century CE. One remnant of the Scythians survived, however, by retreating to the safety of the Caucasus. Their modern descendants, the Ossetians, call 'water' *don*.

Nuristani, the third daughter of Indo-Iranian, is something of a puzzle, because it has only been documented in modern times and so its ancient backstory has had to be reconstructed from texts written much later. The general view is that it falls between the Iranic and Indic branches. As long as it has been recorded, it has been spoken in mountainous parts of Afghanistan and Pakistan, but its original homeland is unknown.

At the moment that they are first made visible in writing, therefore, the Aryans are located in four strongholds: Avestan-speakers in and around Iran, Nuristani-speakers high in the Hindu Kush, Sanskrit-speakers in Syria and, almost simultaneously, at the northwest frontier of India. Any account of the origin and spread of the Indo-Iranian languages ultimately has to explain how they ended up in those places, speaking those languages, when records began.

For many that explanation starts a thousand years earlier, with a chill. The climate crisis of 2200 BCE affected not just Europe but the entire planet. Many regions were hit by drought. The Akkadian and Sumerian Empires expired. People were forced, once again, to move.

At the onset of the crisis, not only were Indo-Europeans to be found throughout Europe, there were as many of them inhabiting the steppes north of the Black and Caspian Seas as those grasslands could support – 'with no leeway except what could be gained by warfare', as one study put it.[7] Such statements are based on evidence of overgrazing as reflected in the state of the soil preserved, like a time capsule, beneath burial mounds. The Yamnaya economic model had become a victim of its own success. People of Corded Ware ancestry, who had once swept all before them as they moved west into Europe, found themselves haunted by famine, perhaps also by plague and salmonella. Some of their descendants turned round and, with their scrawny, dwindling herds, headed back east.

(It's worth dwelling for a moment on the direction of this migration. In western Eurasia, people tend to think of history as having been punctuated by waves of incoming, aggressive nomads from the steppe, from the Scythians to Genghis Khan, but that's because immigrants are more visible than emigrants. Over the *longue durée*, there have been movements in both directions, as the climate oscillated, forests contracted and grasslands expanded, or vice versa.)

As this eastward return gathered steam, it became increasingly competitive and violent: a fight for survival. A new military ethos coalesced, along with new military hardware, so that clans that had already established themselves in the east could come to the defence of relatives or allies further back in the chain. By far the most important innovation in this hardbitten new world was the chariot: a two-wheeled vehicle consisting of a draft pole, two spoked wheels connected by an axle, and a platform. In its simplest form it was pulled by a pair of horses and held two people standing. This was the vehicle that now drew the Bronze Age through the steppe like a curtain.

Archaeologists observe this particular back-migration as a sequence of related cultures that emerge, one out of the other, from west to east, or to be exact, from south-west to north-east – following the rim of the steppe. Geneticists report that the people associated with these cultures were genetically almost identical. From the Dnieper Valley, the migrants followed the forest-steppe belt towards the Volga, spreading along that river and its tributaries towards the Urals. It took about six hundred years altogether, but as the generations passed, more and more of them crowded into the foothills of that mountain chain. Gravitating towards the increasingly rare marshes that herders favour for winter pasture, they became more sedentary – not because they took up farming (they didn't), but because they became fiercely territorial. Compared to the settlements of their Yamnaya ancestors, theirs left easily detectable impressions in the landscape.

Finding themselves in a region rich in metal ores, these herders now began to tap this lucrative resource. As they did so, they were drawn into a trade network for bronze artefacts that had its roots in the Altai to the east and extended all the way west, via the Volga and other rivers, to Scandinavia (the same river network that the Vikings would use later to fetch and trade slaves). The craftspeople who participated in this network took metalwork to new heights, fashioning swords and spear- and axeheads that were coveted far and wide. Among them were the people of the Sintashta culture. From the air a Sintashta town looks a little like a snail shell. Rectangular, semi-subterranean dwellings were arranged in tight concentric circles around a central space, and beyond the outer circle stood earthen ramparts reinforced by timber. The archaeologists who excavated these sites found that each dwelling contained vestiges of ovens, copper and slags. Homes doubled as workshops.

In the last decade, the Sintashta people's genes have given up their secrets. Once thought to have had roots in Asia or the Middle East, it turns out that they were actually related to the people of the Corded Ware culture. They represented, in a genetic sense, the steppe returning to the steppe, albeit after a long interlude in the forests of temperate Europe. Sintashta death rites owed a debt to the Yamnaya, in the form of kurgans and sacrifices of sheep, goats and cattle. Yet in technological and social terms, this culture marked an important break with the past.

Master chariot-builders, who transformed the lumbering wagon into something so swift and light that even gods wanted to race it, they also bred horses on a huge scale. The same study that suggested that the Yamnaya had had a hand in domesticating *Equus caballus* showed that the fully domesticated lineage exploded out of the Sintashta heartlands a little before 2000 BCE. Almost every one of the estimated sixty million domesticated horses presently roaming the planet can trace its origins back to those reared in the rainshadow of the Urals at around that date. The Sintashta embraced a new concept of military leadership too. When a great warrior died, they put a team of horses to death to accompany him to the afterworld, and organised his burial space as if to depict his last chariot ride.[8]

That portrait of the Sintashta, drawn up by archaeologists and geneticists, matches the description that linguists have assembled of the speakers of Proto-Indo-Iranian based on reconstructions from its daughter languages, including Sanskrit and Avestan. The Sintashta are thought to have invented the chariot at around the date that Proto-Indo-Iranian was spoken, four thousand years ago. Sanskrit and Avestan have the same word for a 'chariot' (*ratha*) that is derived from the Proto-Indo-European word for 'wheel', *$roteh_2$*. The Vedas and the *Avesta* share a reverence for chariots ('Thought was her chariot and the sky was its canopy,' says the *Rig Veda*) and for the carpenters who built them. Both sets of scriptures are peppered with such expressions as 'carpenters of words' or 'to carpenter a song of praise'.

Proto-Indo-Iranian was a language of herders, rich in the vocabulary of horses and cattle but poor in that of crops. Among the few farming-related words it contained was **Hajra-*, from Proto-Indo-European **h₂eg'ro-* (the root of '*agri*culture' itself). In the European branches of the Indo-European language family, this word acquired the meaning of a cultivated field, but in Proto-Indo-Iranian it seems to have retained an older meaning, that of a plain into which one drives cattle. That asymmetrical shift in meaning makes sense if one branch of an ancestral population had stayed in Europe, drawing closer to farmers and their food-growing methods, while another had struck back out into the prairies.

One other line of linguistic evidence strengthens the case that Proto-Indo-Iranian was spoken on the steppe: Uralic-speakers borrowed Indo-Iranian words for various metals, as well as words for such things as 'wheel', 'rope' and 'bridge'. Those northern hunter-fishers knew the rivers that irrigate the Urals like the back of their hand. As those waterways became increasingly important conduits for the bronze trade, the likelihood, corroborated by archaeological evidence, is that the Sintashta hired them as boatmen, prospectors and miners. The Uralic-speakers then absorbed words for concepts that they hadn't known previously. Those loans, the oldest of which are about four thousand years old, are hard to explain if Indo-Iranian had traversed the Iranian Plateau far to the south, a thousand years earlier.

The photograph, in black and white, had been taken from a distance to show the vast scale of the mine. Men streamed up and down a cliff-face on ladders, each one carrying a sack equivalent to a good proportion of his own body weight, via a band stretched taut across his forehead. The face of each one was practically touching the behind of the one above, and they were all slick with mud. They looked like worker ants, like one might imagine the Egyptian slaves who were building pyramids at the time the Sintashta were honing the chariot, like a vision of a pre-industrial world that we thought we had left behind. The picture was taken in 1986, at the Serra Pelada gold mine in Brazil. ('Once they have touched gold,' the photographer, Sebastião Salgado, mused years later, 'nobody ever comes back.'[9])

It's this image that comes to my mind when I think about the Kargaly mines in the Urals where, between three and four thousand years ago, men laboured to extract not gold but copper. The sheer scale of Kargaly bewildered archaeologists when they first began to map it out; or, to be precise, when they realised the gigantic scale of its workings *in the Bronze Age*. They knew that Kargaly had been exploited again, millennia later, in the realisation of Catherine the Great's imperial dreams.

The orefields covered five hundred square kilometres, about the size of the modern city of Prague, or Montreal. Across that area, the archaeologists logged thirty thousand Bronze Age pits. To begin with the mines were open-cast, as at Serra Pelada; later it became necessary to tunnel to reach the receding seams. From those seams, men armed with nothing but axes extracted several million tonnes of ore. Out of that ore, using kilns aerated by mechanical bellows, others smelted hundreds of thousands of tonnes of copper. When Kargaly was operating at full capacity, close to four Vatican Cities of forest had to be felled each year to keep the machine stoked. But Kargaly is in the treeless steppe; the first of those forests was over a hundred and sixty kilometres (a hundred miles) away. The bronze end-products were sent hundreds or even thousands of miles away, in all directions. The wagon tracks that by now criss-crossed Eurasia must have seen heavy traffic. In terms of volume of trade, these bronze roads dwarfed the later Silk Roads.

The ores of Kargaly had been known about since the Yamnaya had mined there, but it was only now that they began to be exploited on an industrial scale. Those responsible for this prehistoric industrial revolution were the people of the Srubnaya culture, descendants of the Sintashta, and just as in the later industrial revolution, the individuals at the coalface – or copperface – led miserable lives. Srubnaya miners and metalworkers lived in settlements in the midst of the open-cast workings, cut off from the outside world for long periods of time. One hilltop settlement, Gorny, had no water source nearby, and in that denuded landscape there was no tree cover to provide shade in summer or to break the bitter Siberian winds in winter.

It wasn't only metal production that took place on an industrial scale in Gorny. Animals were slaughtered in mind-boggling numbers. Some were killed for food, but others were sacrificed in strange rituals that linked the extraction of metal to human fertility.[10] We know that the Srubnaya performed other kinds of rituals involving animals, too. At a site not far from the modern city of Samara on the Volga, west of the Urals, David Anthony and Dorcas Brown found the remains of dozens of dogs and at least one wolf that had been sacrificed and eaten. Srubnaya people didn't normally eat canids, so Brown and Anthony think that these were the remains of an initiation ceremony, possibly the ritual that accompanied the 'death' of a young warrior, in the eyes of society, before he was transformed into a 'dog of war'. If they are right, it is the oldest example of such a ritual to have been identified so far, although finds of dog remains at Sintashta sites suggest that it might have even older roots. The association of dogs with death is a recurring theme in Indo-European cultures. The Greek hellhound *Kérberos* finds an exact linguistic counterpart in *Śárvara*, one of the four-eyed dogs that accompanied Yama, god of the Indic underworld.

As the drought eased on the steppe around 1900 BCE, the heirs of Sintashta spilled back into it. To the west of the Urals they were called Srubnaya; to the east, Andronovo.[11] The two cultures had much in common. Besides a possible association between dogs and death, they had inherited the Sintashta's metalworking skills, their knowledge of horse-breeding, and the chariot. But the Srubnaya invested more in growing crops than the Andronovo, and were probably less mobile overall. Both now expanded massively. Srubnaya extended west as far as the Danube, Andronovo headed south into Central Asia and east as far as Siberia's River Yenisei. The steppes were no emptier than they had been a thousand years earlier – if anything they were more populated – and the reason they were able to do so can be summarised in two words: military superiority. The Andronovo left a trail of triumphal graffiti behind them, in the form of chariot-themed rock carvings.

From this point on, archaeologists and geneticists can trace a sequence of related cultures west and east towards the places where the Indic and Iranic languages were first recorded. As so often in the Indo-European story, the picture is patchy, like a magic lantern show, but when combined with the linguistic evidence it makes for a persuasive account of the dispersal of those languages.

The southernmost Andronovo soon came into contact with the fortified oasis cities of Central Asia, including ancient Gonur in what is now the Karakum Desert. The people of Gonur belonged to a settled, agricultural civilisation. They probably spoke a non-Indo-European language, and the Andronovo might have been drawn to them initially as customers for their bronze goods.[12] Geneticists report that the people buried in the cities' central graveyards were predominantly of Iranian farmer ancestry, as might be expected of the indigenous inhabitants, but that by the early second millennium BCE a few individuals buried in outlying cemeteries carried Corded Ware ancestry. Those were likely Andronovo herders.

By 1800 BCE, hundreds of seasonal campsites were clustered around the desert cities. Archaeologists have extracted masses of sherds of Andronovo pottery from these sites, and they have found matching sherds inside the city walls too. There's evidence that the migrant herders were allowed to graze their animals on the stubble of harvested fields, hinting at some give-and-take with the farmers, but there may also have been episodes of violence. The towns began to shrink. As more Andronovo came down from the north, the market for bronze may have reached saturation, triggering new migrations. The future Indic-speakers moved off in two directions, south-west towards Syria and south-east towards India. By 1500 BCE, Sanskrit had been carved into stone in Mitanni lands, and soon afterwards the Vedic poets were at work in the Punjab.

Why one stream of Indic-speakers headed for the Mediterranean is unknown, but in 2007 David Anthony proposed a hypothetical scenario based on the distinctive uses for which the Mitanni reserved Sanskrit. He suggested that a band of mercenaries split from the main Indic-speaking population and headed west at the behest of a Hurrian-speaking king. The mercenaries ended up usurping the crown, not an unusual happening in human affairs, and as the new ruling dynasty they adopted almost every aspect of the local language and culture except for the Sanskrit counterparts they were loath to give up – royal names, the names of deities, and the vocabulary related to horses and charioteering that they had probably brought with them in the first place.

Gradually Iranic- and Indic-speakers lost contact with each other and their languages became mutually unintelligible, but both retained traces of their cohabitation with the people of Gonur. From those settled farmers they took vocabulary relating to irrigation (canal, dug well) and urban architecture, along with words for 'camel', 'donkey' and 'tortoise'.[13] More intriguingly, they seem to have embraced some of Gonur's religious concepts and practices. These included the name of a certain warrior-god, Indra, and his favourite intoxicant *soma*. Or as it was known in Avestan, *haoma*.

The Iranic-speaking community that stayed in Central Asia expanded like a soap bubble again, its dialects diverging into distinct languages. Many of the speakers of those languages would eventually be drawn into the Persian Empire which, by the time Darius died in 486 BCE, controlled about two-thirds of the Eurasian land mass. Among the subject peoples listed on his tomb near Persepolis are the 'haoma-drinking Scythians'. But Darius had had the greatest difficulty bringing the Scythians to heel, Herodotus tells us, because they simply melted away before his armies. In a famous passage, the Greek historian recounts how an exasperated Darius eventually sent a note to the Scythian king, Idanthyrsus, asking him why he kept running away. Idanthyrsus replied:

> If you want to know why I will not fight, I will tell you: in our country there are no towns and no cultivated land; fear of losing which, or seeing it ravaged, might indeed provoke us to hasty battle. If, however, you are determined upon bloodshed with the least possible delay, one thing there is for which we will fight – the tombs of our forefathers. Find those tombs, and try to wreck them, and you will soon know whether or not we are willing to stand up to you.[14]

Without imputing words to people who left us none, it's tempting to think that a Yamnaya or Corded Ware warrior might have expressed a similar sentiment, if challenged, two thousand years earlier.

The migrants came through the Hindu Kush carrying the steppe ancestry that they had inherited from the Corded Ware people of eastern Europe. Some of them decided to settle in the high valleys of Afghanistan, seeding dialects ancestral to today's Nuristani languages. Steppe ancestry has been found in those valleys, in human remains that post-date the exodus from Gonur. Others carried on, descending through the foothills towards the plains of the Punjab. By 1600 BCE steppe ancestry was in Pakistan's Swat Valley, and from there it diffused through the entire subcontinent.

No serious scholar denies that there was a migration into India at this time, but the question that has not been resolved, to everyone's satisfaction, is whether those migrants brought an early form of Sanskrit and the beliefs that shaped the Vedas. The steppe theorists, who trace the Indo-Iranian languages back through the Andronovo to Sintashta, say that they did. Those who think that the Indo-Iranian languages crossed the Iranian Plateau two thousand years earlier disagree. For them, Sanskrit was *already* spoken in India when those migrants arrived in the north. And there is a third theory, which is that the migrants did not bring Sanskrit because Sanskrit was born in India, being the language of the Indus Valley Civilisation.

This last theory, dubbed 'Out-of-India', reverses the arrow of Indo-European dispersal and places the original homeland of the entire family in north-west India. It is supported by a very small minority of scholars, but it's important to mention it for two reasons. First, it draws attention to a genuine weakness in the steppe hypothesis, which is its assumption that genes and language travelled together. Second, it enjoys a large following among non-scholars in India, where the debate over Indo-European origins has become so politicised that each new scientific finding is regarded, by one group or other, as ideologically suspect. At different times in different places, the Indo-European story has been beaten like warm copper to fit a political mould. It happened when British imperialists wanted to justify their colonisation of India (they were simply retracing Aryan steps). It happened when the Nazis needed a precedent for *Lebensraum* (they were merely reclaiming their Aryan birthright). Today, the story is still being beaten out of shape in Europe (and America), but the beating is most energetic on the subcontinent, where Out-of-India has become dogma for many Hindu nationalists.

In 2001, before studies of ancient DNA took off, a British-born American Sanskrit expert named Edwin Bryant collated evidence for the origins of the Indic languages in a way that was widely praised for its balance. Importantly, he included Indian scholarship on the question, noting that Indo-European studies had been dominated by Western academics who had too often ignored the expertise of their Indian counterparts. Indian scholars have long insisted, for example, that the Vedas are a collection of *myths*. Those myths might have been inspired by real life, but they were never intended to be a faithful account of history. Besides, Westerners who had mined them for possible references to actual events had cherry-picked the evidence, conveniently overlooking the fact that they make no mention of an Aryan invasion. Though it was likely that the oldest parts of the *Rig Veda* were written down around 1500 BCE, Bryant concluded, it was not impossible that they were a thousand years older, or that the language in which they were written was older still.

What has changed since Bryant's book, *The Quest for the Origins of Vedic Culture*, is that we now have evidence that immigrants reached northern India starting around 1600 BCE or a little earlier, and that they brought steppe ancestry with them. The Harappan settlements – hundreds of them in all, including at least five cities – had risen from the Indus floodplain and neighbouring regions a thousand years before that, and flourished for five hundred years. Vanishingly little Harappan DNA has been analysed, because conditions on those plains aren't conducive to its preservation, but the little that has suggests that the Harappans had no genetic connection to the steppe. For aeons they had been indigenous to the subcontinent, and they had likely invented agriculture independently of developments in the Near East – a point against the theory that Indo-Iranian languages had arrived with farming, and farmers, from the west.[15] It is also true that groups of Indians who speak languages descended from Sanskrit today typically carry more steppe ancestry than those who speak non-Indo-European languages. And that the traditional guardians of the holy texts, the Brahmins, have more steppe ancestry than other social groups. There is an indirect connection between Sanskrit, the Vedas and steppe ancestry, in other words. We have no *direct* evidence that the ancient immigrants spoke Sanskrit, however, and there are plenty of examples of intruders switching to the language of the natives – not least the Scythians, Greeks and Mughals who invaded India later. It is therefore at least possible that the immigrants embraced the language of the Harappans, taking up Sanskrit upon their arrival in the Punjab.

The only hope of settling the debate definitively is for someone to decipher the Harappan script. This has obstinately resisted all efforts to crack it for a century, not least because no Harappan equivalent of the Rosetta Stone has been unearthed – no bilingual document that encodes their language and another known language that would allow a comparison of the two.[16] Such a document could still come to light, since only a fraction of Harappan settlements have been excavated, but for now their tongue eludes us.

To date, then, the steppe hypothesis offers the best account of the evidence. At no point since Oriental Jones' address to the Asiatick Society in 1786 has anybody questioned the Indo-European nature of Sanskrit, so in the final analysis there are really only two possibilities: either it came to India from outside, or India is the homeland of the entire family. Proponents of the Out-of-India theory have made very little effort to explain how the other Indo-European languages ended up where they did. So bearing in mind Bryant's important caveats, let's go forward on the basis that the migrants heading through the mountains between three and four thousand years ago – perhaps via the Khyber Pass, like so many who came after them – spoke a form of Sanskrit.

The Arya seem to have come in families, since both sexes are represented among the human remains that have been sampled. They were still herders, if the *Rig Veda* is anything to go by, and still given to cattle-raiding, but they may not have been alone. The Indic languages contain a layer of non-Indo-European loanwords that they don't share with the Iranic ones, meaning that the loans must have been absorbed after the two branches split. But the source of those loans looks to be the same one that, earlier on, lent both branches Indra's name and the cult of soma: the language of Gonur. It seems possible, then, that some of those farmers quit their oasis (perhaps because the desert was encroaching?) and threw in their lot with the eastward-migrating Indo-Europeans. They may even have taught their travelling companions some skills along the way. The loans include *pushpa* (flower), *pippala* (sweet fruit) and *urvaruka* (cucumber), as well as words for 'granary' and 'root'. By the time the Aryans arrived in India, they may finally have learned the basics of farming.

As they descended into the plains, they would have caught their first glimpse of the Harappan cities, laid out on their mesmerisingly regular grids, with their brick houses and many baths. In the *Rig Veda*, the arch enemies of the Aryans were the Dasyus, people who upset the cosmic order because they failed to perform the correct rituals to the correct gods using the correct form of words. ('Aryan' and 'Dasyu' were cultural and not ethnic categories, in other words, though the two groups may also have spoken different languages.) Indra rode out in his chariot and defeated the Dasyus, destroying their fortresses and securing their land and water for his people. Some have speculated that those heroic tales were inspired by the actual destruction that the Aryans inflicted on Harappan cities. But there are no signs of conquest in those cities, no layer of burnt destruction as there is at Hattusha or Persepolis. What evidence there is speaks more insistently to an advanced civilisation dying slowly as the monsoon failed – probably an effect of the same climate crisis that pushed Corded Ware people back to the steppe – and its inhabitants ebbed away.

The hymns of the *Rig Veda* were already old by the time they were written down. Even if Indra's exploits were inspired by real events, the poets may have been recalling earlier confrontations, such as those between the Andronovo and the people of Gonur. The Aryans are more likely to have encountered ghost cities in the

Indus Valley, or at least much-diminished ones – as the *Rig Veda* implies. They may even have settled among them (planting sweet fruit and cucumbers in the interstices). In the 1970s, an American archaeologist called Gregory Possehl remarked upon the extraordinary 'empty spaces' between the Harappan settlements, which he felt must have been occupied when those settlements were still inhabited. Though little usable DNA has been extracted from the Indus Valley, human skeletal remains have been found there, and they speak to a variety of anatomical types. The foreigners, who would still have stood out by virtue of their size and colouring – and certainly by their language – may have mingled with the last denizens of an old and very sophisticated civilisation, individuals who wouldn't or couldn't leave and who were occasionally spotted flitting through its crumbling streets.

By 1500 BCE, when it first appeared in writing, Sanskrit was itself a cabinet of curiosities. It had been influenced by all kinds of other languages, only some of which have been identified, and those influences have been inherited by its offspring. One is a type of consonant called a retroflex, because it is formed by curling the tongue back against the roof of the mouth. Hindi, for example, does not have the English consonants *d* or *t*, but it does have the retroflexes $\d$ and $\t$. When speaking English, native Hindi-speakers often substitute these for their nearest English counterparts.

Retroflexes are rare outside the subcontinent, but they are found in the Dravidian and Munda language families.[17] Dravidian is now spoken mostly in southern India, while Munda can be heard in pockets scattered around the north-east of the country, but some scholars suspect that one or other was the language of the Indus Valley in the Bronze Age. There is again no way of proving this until the Harappan script is deciphered, but the suggestion is that, in taking up the tongue of the Indo-European-speaking immigrants, Harappans were unable to rid themselves of the retroflex which consequently became a fixture in Sanskrit. When they drifted away in search of greener pastures, they left the retroflex behind as their calling card. Sanskrit, like all languages worthy of the name, is a hybrid that reflects its unique journey through the world. An auspicious hybrid, in its case, charming like a creeper.

In the late spring of 1891, peat-cutters working a Danish bog happened upon a silver bowl. In it were some loose silver plates, either convex or concave, that had once been fitted to its inner and outer surfaces. The plates had been beaten into images that told a story, but what that story was, nobody then or now could divine. All they could say was that it involved war, sacrifice, male and female deities and an eye-catching figure sporting a pair of stag's antlers. In one hand the figure gripped a ram-headed snake, in the other a torque.

The object became known as the Gundestrup cauldron, and from the moment the Danish government offered a reward to its discoverers it caused ructions. First the peat-cutters squabbled over the reward, then scholars disputed its provenance and much else besides. Some facts have been established, more or less, in the last hundred and thirty years. The great bowl was probably filled with drink served up ritually to warriors before a battle, or to celebrate their victory on their return. It was probably made in the century or two before the Common Era, by no fewer than five top-flight silversmiths. The smiths were not local, since there was no tradition of skilled silverwork in northern Europe at that time. It had passed through many hands and been much repaired, but the repairs were not up to the standard of the original work. It may have been placed in the bog as an offering, prehistoric people having regarded bogs as sacred places, or it may have been stashed there for safe-keeping, perhaps in a time of war. Where it was made, and for whom, are unknown to this day.

The imagery on the cauldron is often described as Celtic. The figure with the antlers has been identified as the Celtic god of wild things, Cernunnos or 'the horned one', and the Celtic war trumpet, the carnyx, is very much in evidence. But many of the motifs would have been familiar to Germani, Scythians and Indians too, from the wheel gods to the slaying of a bull to the ritual baptism of wolfish combatants. Elephants are depicted, badly (they are no doubt copies of copies, executed by people who had never seen one in the flesh), and there is a goddess with plaits and a pair of birds who, according to the British archaeologist Timothy Taylor, would have been recognisable to Indians as Hariti and to Welsh people as Rhiannon. The silverwork is probably Thracian, from the region west of the Black Sea that now falls between Bulgaria, Türkiye and Greece. Taylor and his Swedish colleague Anders Bergquist have proposed that the cauldron was made in the neighbourhood of Transylvania (in modern Romania) when Thracians, Celts and Germani were fighting over that region.[18] A Celtic chieftain might have commissioned it from local craftsmen and a Germanic tribe might have seized it and carried it back to Jutland as loot.

Neither the story of the cauldron nor the story the cauldron tells is ever likely to be elucidated in full, but it is relevant to our tale because of the snapshot it provides of the Indo-European world at the instant the bog swallowed it up. Though Aryans, Romans and Celts saw each other as barbarians, though they spoke mutually unintelligible languages and worshipped different gods, their psychic universes overlapped. From the Scottish isles to the Himalaya, there existed a chain of societies that was deeply interconnected through trade, custom, language and mythology.[19]

The *ashvamedha*, the Indian horse sacrifice, finds echoes in ancient Rome and Ulster (and, I contend, in the calumny concerning Catherine the Great, according to which the Russian empress died while attempting to have sex with a stallion). In a ritual called October Equus, the Romans held a chariot race on the ides of October – the fifteenth of that month – and sacrificed the right-hand horse of the winning pair to Mars, god of war. The animal was speared to death and again dismembered. The Irish added their own twist: it wasn't the queen who had sex with a stallion, in Ireland's royal capitals, but the king who copulated with a mare. The mare was then cut into pieces and boiled to make a broth, and the king bathed naked in the broth while chewing on hunks of its flesh.

The similarities between these traditions suggest that early Indo-Europeans had a ritual for the renewal of kingship that involved a king or queen having, or simulating, intercourse with a horse which was then sacrificed. The royal coupling may have led to the birth of divine twins, in the associated mythology, since there is a strong link between twin deities and horses in Indo-European cosmogony – from the Greek horsemen Castor and Pollux, immortalised as the constellation Gemini, to the Ashvins, the twin charioteers of the *Rig Veda*, and Horsa and Hengist ('Horse' and 'Stallion' in Old English) who, legend has it, led the Anglo-Saxons to Britain.

Myth-making and storytelling were intimately bound up with sacrifice in the Indo-European world. They were core elements of the rituals that maintained order in those societies. 'Narrative emerged during the wait, the long wait for the horse's return,' wrote the Italian classicist Roberto Calasso of the *ashvamedha*. 'And it was a way of preventing the relationship with the wandering horse from being broken.'[20]

The ones who crafted the narrative, who prevented that relationship from being broken, were the poets. To become a poet demanded long years of training, and as with smithery or carpentry the skills were often passed down through families. On the Welsh island of Anglesey, as in the Persian city of Shiraz, bards memorised large bodies of poetry and incantations and learned the metrical rules of composition, prosody and more. They were recognised as skilled artisans, and richly rewarded by the nobles in whose courts they served.

Each time a poet performed, they adapted their song to their audience, combining old motifs in new ways and sometimes introducing new ones, so that the song evolved but it also stayed the same. Among the oldest songs that Indo-European poets sang, mythologists tell us, is one about a smith who makes a pact with a devil. Having swapped his soul for the power to weld any materials together, the smith cunningly welds the devil to an immovable object. The Brothers Grimm collected variations on this theme in rural Germany in the nineteenth century. It is the kernel of Faust, a story reworked by Marlowe, Goethe and Mann. If the mythologists are right, however, it is at least six thousand years old – old enough to have been inspired by the smiths at Varna, those alchemists who, wreathed in flame and acrid smoke, transformed rock into gold.

Writing is often described as a miracle, and it certainly made many things possible. If you think about it, however, the real miracle was that a story about a smith and a devil was handed down, more or less faithfully, for thousands of years *before* it was scored into a durable material. By repeating the old familiar stories, with variations, the poets stitched the community together across generations and across continents. They narrated the events that were drummed into the Gundestrup cauldron and into the Indo-European psyche. It was the poets who ensured that, though we may not be able to place the plates in the cauldron in the right order, any Greek, Thracian or Aryan could have done.

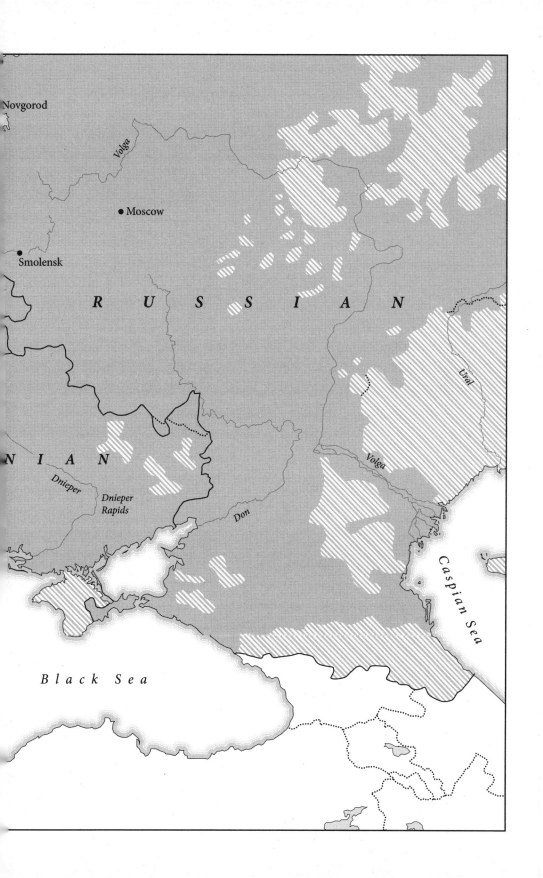

7

Northern Idyll

Baltic and Slavic

In 1962, almost twenty years after fleeing her native Lithuania, Marija Gimbutas published a book called *The Balts*. It was a distillation of her encyclopaedic knowledge of the Baltic region, from its folklore and religions to its languages and archaeology, but she couldn't quite keep her nostalgia out of it. She was, after all, an exile, having been forced to stay away from Lithuania by its post-war absorption into the Soviet Union. Sitting atop a Californian hill, with broad vistas in all directions, she saw in her mind's eye the hills blanketed with oaks that surrounded Vilnius. 'The Californian sand dunes, at Carmel, remind me of the pure white sands of Palanga, where I used to collect handfuls of amber; and the sunsets in the Pacific, of the peacefully sinking sun as it disappeared into the Baltic Sea, beyond where, to the west, my forefathers thought was the cosmic tree, the axis of the world, holding up the arch of the sky.'[1]

The Baltic languages are spoken today in Lithuania and Latvia, while the dominant languages in neighbouring Estonia and Finland belong to the Uralic family. Baltic wasn't always spoken in those places, however. As with all languages, its territory has shifted over time. Gimbutas understood that the story of Baltic was intimately entwined with that of its neighbours in the region – Germanic, Slavic and Uralic – but she couldn't help thinking of it as a case apart. In her mind, the Balts and their languages had stayed closer to their roots by virtue of their isolation in a land of forests and lakes, far from the highways of ancient Eurasia. She pictured her countryfolk doing the same things they had done for centuries, if not millennia: tilling the earth, living in harmony with the seasons, singing as they worked: 'The Baltic area is exceptional among the lands inhabited by people of Indo-European origin in that the language and folklore have survived in a remarkably pure state . . .'

She wasn't the only one to express this thought. It was received wisdom, in 1962, that Lithuanian in particular was a frozen vestige of Proto-Indo-European. The idea went back a long way. 'Whole Sanskrit phrases are well understood by the peasants of the banks of Niemen,' stated an 1882 edition of the *Encyclopaedia Britannica*, referring to the river that rises in Belarus and traverses Lithuania on its way to the Baltic Sea. 'Anyone wishing to hear how Indo-Europeans spoke should go and listen to a Lithuanian peasant,' wrote the French linguist Antoine Meillet in 1913. And it's still often said, both inside and outside academia, that Lithuanian is the living Indo-European language that most closely resembles that long-dead ancestor. What was not to intrigue a young historical linguist searching for a specialisation in the late 2010s? But as soon as Anthony Jakob started looking under the hood of the Baltic languages, he realised that the truth was both more nuanced and more interesting.

Lithuanian *sounds* archaic, Jakob says, in part because of a quirk it shares with its extinct relatives, Ancient Greek, Latin and Sanskrit, which is that many of its nouns end in *s* in the singular. The Lithuanian word for 'man', *výras*, is very similar to its Sanskrit counterpart *vīrás*, while Lithuanian *ugnìs* (fire) resembles Sanskrit *agnìs* ('man' and 'fire' have become *bīr* and *āg* respectively in Hindi).

But in other ways Lithuanian has evolved quite as much as any of its relatives. It has updated the grammar of Proto-Indo-European by, among other things, laying off a number of verb tenses. And its vocabulary has seen a comparable rate of turnover, accumulating substitutions as is normal in the process of language evolution. The Lithuanian word *galva* (head) is unrelated to its Proto-Indo-European counterpart *$k'erh_2$*- (the root of Hindi *sir*, Persian *sar* and English *cerebral*), and although Lithuanian has a word, *esti*, that is related to French *est* and English *is* (the third person singular of the verb 'to be'), Lithuanians no longer use it. They say *yra* instead, while Latvians say *ir*. Nobody knows where these two words come from.

In fact, Lithuanian adheres strictly to the rule that there is no such thing as a pure language. Its peculiar mix of archaisms and innovations reflects the periods of calm that it has known, as well as the periods of upheaval, and the other languages that it has brushed with along the way. Jakob, who defended his doctoral thesis in 2023, is one of the more recent recruits to the effort to piece that backstory together from the linguistic evidence. The result has been a sea change in thinking about the origins of Baltic and Slavic.

In Meillet's day, and still in Gimbutas' day, Baltic and Slavic were considered sufficiently different that they must always have been distinct branches of the Indo-European family. Now, armed with more information and finer resolution, most linguists acknowledge that the differences between them were racked up relatively recently. Baltic and Slavic were once the same language, they believe, and even after they split remained close for two thousand years or more. Events triggered by the fall of the Western Roman Empire caused Slavic to barrel away from its more conservative sister, but Baltic didn't stay the same, and the linguistic legacy of that rapid divergence masked the long period of stability that preceded it. It was that recent divergence that had led Meillet and others astray.

The realisation that Lithuanian isn't the fossil it was once thought to be has solved some mysteries but accentuated others. If it has followed its own path, why does it have so much in common with Sanskrit? The similarities are there in the other extant Baltic language, Latvian, and in the Slavic languages too, but they are most pronounced in Gimbutas' mother tongue, and they extend from sound to vocabulary and even mythology. The divine twins are called *Ashvins* in Sanskrit and *Ashvieniai* in Lithuanian, and both names evoke horses.

A seemingly outlandish theory, that Balto-Slavic and Indo-Iranian were, in turn, once the same language, has been gaining currency of late. The theory isn't new, but it has always been controversial, and it's not hard to see why. An ancient nexus between languages that would eventually be spoken so far apart – one among the ruined metropolises of the Indus Valley, the other on the amber-strewn beaches of the Baltic – strikes many as beyond the bounds of probability. And yet in the last few years archaeologists and geneticists have produced evidence that such a nexus really existed. If you pull on the cultural and genetic cords that tie the speakers of Sanskrit and Lithuanian to their roots, you find that they cross in a single place and time: the Dnieper Rapids, in the Bronze Age.

By 2800 BCE, the people of the Corded Ware culture had already scorched a path into central and northern Europe. In the east they had gravitated like so many of their ancestors to the Middle Dnieper, south of modern Kyiv. There, where the water roared over a series of rocky shelves, they supplemented their diet of milk and meat with freshwater fish.

For hundreds of years, all their needs were satisfied in this land of plenty. Starting around 2400 BCE, however, the first augurs of the climate crisis would have made themselves felt, and as the colder, drier conditions intensified, and it became harder to eke out a living, some raised their eyes to the east. The migration that now got underway culminated in the Sintashta culture, the presumed speakers of Proto-Indo-Iranian who lived south-east of the Urals. But it began with the herder-fishers of the Middle Dnieper, who set off towards the Volga and the mountains beyond in search of fresh pasture for their bony herds.

The Sintashta were expert metalworkers and chariot-builders, but fundamentally they were herders. They formed the eastern extremity of a cultural continuum that began with the Corded Ware clans on the Dnieper. Genetically, too, the two groups were related. These findings have emboldened some linguists to claim that laid over the cultural and genetic continua was a linguistic one: a spectrum of dialects that bridged Balto-Slavic and Indo-Iranian. Eventually these would split into two branches that would go their separate ways (and each would then split again, and again), but for a while, people living throughout that wide space could understand each other. It's a mind-bending thought, and most historical linguists do not (yet) subscribe to it. But a growing number do, and the reason is that a dialect continuum across the western steppe could explain those puzzling similarities between Balto-Slavic and Indo-Iranian.

Of the features that Balto-Slavic and Indo-Iranian share, the ones that linguists consider most revealing in terms of reconstructing the two branches' past are satemisation and another sound change called the 'ruki rule'. Under the ruki rule, an *s* inherited from Proto-Indo-European was systematically transformed into a sound resembling English *sh* after *r*, *u*, *k* or *i*. (An example is Proto-Indo-European *$h_2eusōs$, meaning 'dawn', which became *austrōn-* in Proto-Germanic – the root of English 'Easter' – but something closer to *aushra* and *ushas* in Lithuanian and Sanskrit respectively.)

These two sound changes affected both Balto-Slavic and Indo-Iranian (and in the case of satemisation, some other branches of the family as well), and they are thought to have taken place at roughly the same time. Some linguists consider the ruki rule to be so specific that – like some fluke genetic mutation – it can only have arisen once, and that Balto-Slavic and Indo-Iranian must therefore have been a single language when it happened. Others argue that it could have come about independently in two distinct branches. You can see why so much rides on this question, and why the ruki rule is so hotly debated. If Balto-Slavic and Indo-Iranian had to have been one language when they were 'rukied', Indo-Iranian must have crossed the steppe to reach the parts of Asia where its descendants are spoken today. If not, it could have left Anatolia with Neolithic farmers, passing south of the Caspian Sea and never coming within earshot of Balto-Slavic. On the ruki rule rests the origin story of the entire Indo-Iranian diaspora.

That there was once a nexus between Baltic and Slavic is far less controversial, given how long they have been neighbours, and is now widely accepted. Many of the tributaries of the Dnieper in its middle and upper reaches – in modern Ukraine, Belarus and Russia – have names that linguists deem to be Baltic or Balto-Slavic in origin.[2] And Baltic and Slavic share a fairly rich body of vocabulary connected with rivers and fishing. Latvian and Lithuanian, spoken to the east of the River Niemen, once had a sister called Old Prussian that was spoken to the west of that river. The Baltic language Old Prussian had a word for 'sturgeon' (*esketres*) that is related to the name for the same fish in Russian, a Slavic language (*osëtr*). Lithuanian (Baltic) *šămas* and Russian (Slavic) *som* spring from a single root meaning 'catfish'. And Baltic and Slavic share words for 'wading', 'diving', 'spawning', 'dugout canoe' and 'raft' that have no counterparts in any other branches of the Indo-European family.

Linguists infer that Baltic and Slavic were fused at the time that they invented or borrowed this riverine vocabulary. They cannot have inherited it from Proto-Indo-European, because Proto-Indo-European lacked it. And since Baltic and Slavic have almost no agricultural vocabulary in common, the speakers of Balto-Slavic are unlikely to have been farmers. Their lexicon is a snug fit for the herder-fishers of the Corded Ware culture who archaeologists say were living along the middle stretch of the Dnieper in the early third millennium BCE.

Balto-Slavic probably remained a single language until about 2000 BCE, when some of its speakers started to move north and east, towards what is now Moscow and the upper reaches of the Volga. As discussed, there are scholars who think they were following in the footsteps of the Indo-Iranian speakers to whom they were also related genetically, but these later migrants spoke a forerunner of Baltic.[3] The linguists who meticulously surveyed the river and place names of central and eastern Europe, in the early twentieth century, tentatively mapped the prehistoric range of the Balts as stretching from Poland to central Russia – as far east as the River Oka, a tributary of the Volga. On that basis, Gimbutas wrote, the Balts were misnamed. They had once lived far from the Baltic Sea, and before the expansion of the Slavs and Germani out of their respective homelands had inhabited an area six times the size of their modern territory. But she was writing before linguists recognised an initial Balto-Slavic phase of the language. It seems more likely now that Balts and Slavs named some of those rivers jointly, before they saw each other as Other. Slavic may have been the language that developed among those who stayed behind once the future Balts had left. Some of them followed the Dniester towards the Carpathian Mountains and present-day Poland, but eventually, as we'll see, they would spread much further afield. The genetic evidence, where it is available, corroborates this scenario. The diversity of Y chromosomes among modern Slav men can be explained by a small, initial cluster that radiated out of the Middle Dnieper Basin.

Though they followed different trajectories, both Balts and Slavs now entered the forests that, four thousand years ago, covered all of northern and central Europe. In doing so, they encroached on farmers' territory. From the villagers with whom they now entered into dialogue, they received their primary schooling in woodland living. They learned to haggle in beans, carrots and turnips, and to point out elks, woodpeckers and hawks to each other. We know that they had parted ways by the time this happened, because they borrowed different words for the same species. And it wasn't just their vocabulary that had begun to diverge. Baltic and Slavic were obeying different sound laws by now. They had become distinct languages. They would remain close for a very long time, but they wouldn't meet again for another two thousand years.

The Proto-Slavic homeland is rather fuzzily defined, especially in the west, but most scholars locate it roughly between the headwaters of the Dniester and the Middle Dnieper. It fell within the borders of modern Ukraine, in other words, close to the frontier with Poland. From the eighth century BCE, the Slavs' neighbours on the prairies to the south and east of them were the brilliant, bejewelled Scythians, those Iranic-speaking heirs of the Sintashta who outsmarted Darius, for a time, and who managed to impose their own names on many of the region's most significant waterways – that drumbeat of Ds.

In the early centuries of the Common Era, the East Germanic-speaking Goths came through the Slavic heartland, on their way from the Baltic island of Gotland to the Black Sea. A legacy of this meeting is the Russian word for a parliament, *duma*, which Gothic loaned to Slavic. (The original Gothic word was *dōms*, meaning 'judgement'.) By the fifth century the Roman Empire had fallen in the West, and Europe had entered a period of turmoil which German-speakers call the *Völkerwanderung* (wandering of peoples). The climate cooled again. Wars and dangerous diseases ravaged the continent. It was a time of danger and opportunity, and the Slavs took to the highways.

They did so initially as part of a military federation ruled by Huns and Avars, Turkic-speaking horsemen from the steppes. Some went west, crossing the Elbe into German-speaking lands, but the majority followed the Goths south-east into the Balkans. If the stick driving them was the cold weather, the carrot luring them on may have been the lands vacated by Justinian's Plague. There are souvenirs of these campaigns embedded in their language; postcards from the edge. From Old High German in the west, for example, they borrowed *król*, meaning 'king'. This word, which is clearly unrelated to Latin *rēx*, Sanskrit *rāj-* or Gaulish *rīx*, is a corruption of the name of the Frankish king *Kar(a)l*, better known to English-speakers as Charlemagne.

The transformation of Slavic probably began with the attempts of Slavic combatants to emulate the speech of their Avar commanders. (The Avars might have lent the Slavs a word meaning 'gathering place', *maidan* in Ukrainian, that they in turn had borrowed from Iranic.) It continued as the people the Slavs conquered attempted to emulate them. One of the most important changes the language underwent, during this period of upheaval, was something that linguists refer to as the 'open-syllable conspiracy'. An open syllable is one that ends in a vowel; a closed one ends in a consonant (think 'go' versus 'god'). Most languages tolerate both, but during the *Völkerwanderung* the Slavic languages gradually ejected all closed syllables. In the earliest documented Slavic language, Old Church Slavonic, the word for 'head' that we met previously in its Lithuanian form, *galva*, became *glava*. A letter switch turned the closed syllable *gal-* into the open one *gla-*.

Lithuanian *ranka* (hand) became *rǫka*, by folding the *n* into the nasalised vowel *ǫ*, and Lithuanian *ragas* (horn) became *rogŭ*, simply by dropping the final consonant.

By a variety of tricks, in other words, Slavic tore itself wide open. And though that transformation began with Proto-Slavic, it continued even after the proto-language had begun to fragment into its daughters. Hence the label 'conspiracy', because it looks as if there had been a conscious and sustained effort to strip out every last closed syllable from Slavic. In reality, since it happened over many generations, and since people are usually unaware of how sounds are changing in their mouths, it cannot possibly have been conscious, and linguists have long puzzled over what caused it.

One theory, put forward by linguist Eugen Hill, has to do with the fact that all languages prohibit certain combinations of consonants. Modern English doesn't allow *dv*, for instance, while Czech does (think of the composer Dvořák). Nor does English allow *kn*, today, but the spelling of words like 'knife' indicates that it did in the past – and that both letters were once pronounced. At the time of the Slav migrations, most of the Balkan peninsula apart from Greece was Italic-speaking – having previously been claimed by the Romans – but the Slavs would have encountered pockets of East Germanic-speakers too. Italic forbade the combination *dl*, then, while Germanic barred *kt* and *pt*. As the Slavs conquered speakers of these languages, the vanquished started speaking Slavic, but they stumbled over words containing consonant combinations that were prohibited in their mother tongues. (Psychologists report that people don't even hear the combinations that are forbidden in their native language; they hear the nearest pairing that's allowed, which they then reproduce.) The result was that Slavic absorbed prohibitions from a number of foreign languages which, combined, forced out its closed syllables.

Hill's theory is pretty new, and at the time of writing his academic peers have yet to pass judgement on it, but there is some historical evidence to back it up. The Slavs are reported to have been remarkably open-minded towards those they subjugated, often accepting them as full and equal members of Slav society. This was unusual. The Anglo-Saxons, reaching Britain at about the same time, were more inclined to enslave the Celtic-speaking natives. But the Slavs' openness to foreigners and their oddities of speech might have expedited the open-syllable conspiracy – or compromise. Slavic spun away from Baltic and simultaneously diverged three ways, forming western (including Polish and Czech), southern (such as Serbo-Croatian and Bulgarian) and eastern (Ukrainian, Russian and Belarusian) sub-branches.

By the end of the sixth century, according to Roman chroniclers, Slavs and Avars were busy laying siege to Thessaloniki, a port on the Aegean and the second most important city in the Eastern Roman Empire after Constantinople. The Thessalonians spoke mainly Greek, but when the city fell they learned to speak Slavic, probably as a second language to begin with. It was therefore not so strange when, receiving a request from a Moravian prince to send missionaries to spread Christ's teachings among his people in the ninth century, the Emperor and the Patriarch of Constantinople looked for candidates among the Thessalonians. The patriarch selected two brothers, Constantine and Michael, for the mission to enlighten Slavs in what are now the Czech Republic and Slovakia.

Though the brothers spoke Slavic, all religious texts existed in other languages at the time, and an important part of their mission was to translate them. The problem they faced was that the Greek and Roman alphabets were not well suited to Slavic, meaning that they failed to capture some sounds while including others that Slavic didn't use. Constantine, the younger and more intellectual of the brothers, set about adapting the Greek alphabet so that he could more accurately transcribe the Slavic dialect that he and Michael had grown up with. Since the first documents to record

Slavic were the brothers' Bible translations, that dialect, long dead except as a liturgical language, is known as Old Church Slavonic. The brothers are better known as Saints Cyril and Methodius, and the alphabet that Constantine devised would evolve into Cyrillic. (Roman Catholicism had replaced the Orthodox Church in Moravia before the ninth century was out, along with the Roman alphabet, but the Cyrillic script came to dominate further east.)

The western and southern branches of Slavic lost touch after the arrival of the Magyars from the late ninth century drove a Uralic-speaking wedge between them, in Hungary. The expansion of an Italic language, Romanian, helped push apart the southern and eastern branches, and in time the three Slavic-speaking populations ceased to be able to understand each other. As for their shared name, it's fairly common knowledge that 'Slav' is linked to English 'slave', but that obviously wasn't its original meaning (what people would boast of being 'the unfree'?). Some linguists see in 'Slav' a derivation of Proto-Indo-European *kléwos*, meaning 'fame', in which case Slavs may have called themselves something like 'the famous people'.

Starting in the Middle Ages, when Franks took Slavs captive at the borders of the Holy Roman Empire, the Slav name – *sclavus* in the official language of that empire, Latin – became associated with servitude. (You may recall that Latin already had a word for a domestic slave, *famulus*; new times and new categories called for new vocabulary.) Latin *sclavus* morphed into French *esclave*, which became 'slave' when the Normans took it to Britain. But the Slavs themselves retained the first meaning of the word, as witnessed by the Russian boy's name Mstislav (vengeful fame), the Polish one Stanisław (he who has achieved fame), and the Ukrainian war cry *Slava Ukraini!* (Glory to Ukraine!)

NORTHERN IDYLL

At some point in the Balts' migration towards the eponymous sea, from what is now Ukraine, a group broke away and headed for lands west of the Niemen (roughly the modern Russian exclave of Kaliningrad). Their language became Old Prussian which, despite the misleading inclusion of 'Prussia', was a Baltic and not a Germanic tongue. The remaining Baltic-speakers eventually reached the forests north of modern Moscow, where they encountered barbarian hunters sporting sumptuous furs.

The barbarians' language was Uralic – Proto-Finnic, to be precise, the ancestor of modern Finnish and Estonian.[4] They were the bow wave of the western expansion of the Uralic languages out of their homeland, which is now thought to have been located – not in the Urals as their name would suggest – but closer to Siberia. (Proto-Saami, a sister of Proto-Finnic, had already departed towards Lapland, where it would become the language of reindeer-herders.) The meeting of the Proto-Balts and the Proto-Finns unfolded in an unexpected way.

Fortified villages began to spread across the region east of the Baltic Sea, as far north as Finland and as far east as Moscow. You can see their remains today, jutting out of riverbanks and lakeshores. The inhabitants of these forts obviously feared someone; they built ramparts of stone or baked clay several metres high, and reinforced them with timber to keep the enemy out. But the enemy may just have been the inhabitants of the next fort along. It wasn't that the Balts feared the Finns or vice versa, because the two groups cohabited peacefully within.

Distinct pottery types and burial rites have been found in the settlements, which suggest to Estonian archaeologist Valter Lang that Balts and Finns found a way of living together while maintaining their ethnic and cultural differences. Linguists, too, suspect that they formed mixed bilingual communities in which they preserved their mother tongues but loaned words to each other. The Balts were fully fledged farmers by now. The Finns were forest specialists who hunted game and brought back valuable skins and furs. Through a process of trial and error, they may have discovered that they were more prosperous together than apart. The Finnish words for 'seed', 'pea' and 'chaff' are all Baltic loans (Lithuanian/Finnish *sėmuõ/siemen*, *žiřnis/herne* and *pēlūs/pelut* respectively). And during this period of cohabitation, both languages absorbed loans from an unknown third. Since many of the loans relate to fishing – 'fish weir', 'crayfish', 'whitefish', 'eel' – the source may have been a language spoken by hunter-fishers who had inhabited the region long before either Balts or Finns arrived.

> How close you are
> My forefathers!
> They herded cows
> And saddled horses,
> Planted children
> And peas.[5]

The Balts and Finns had been living this communal life for more than a thousand years when the Slavs showed up in their midst. Their language had once been indistinguishable from Baltic, and in terms of vocabulary and to some extent grammar the two were still close. But thanks in large part to the open-syllable conspiracy, Slavic now *sounded* quite different from its sister. The Slavs' wanderings, in which the Balts had not participated, would have been audible. They settled among the Balts and Finns, and by the ninth century speakers of all three languages – two Indo-European and one Uralic – were mingling in the north-western Russian city of Novgorod, close to the borders with modern Finland and Estonia.

Novgorod was a strategic hub on the river network that connected the Baltic south via the Dnieper to the Black Sea and Constantinople, and east via the Volga to the Caspian and Baghdad, and the Slavs weren't the only ones drawn to it. Vikings had crossed the Baltic from Sweden at about the same time, and established a trading

NORTHERN IDYLL

colony there. The Slavs may have carved out some indispensable role for themselves in the city's commercial life, since the language of trade there was an East Slavic dialect that linguists call Old Novgorod. We know this because the city's inhabitants wrote to each other (some of their letters, scrawled in Cyrillic on birch bark, are on display in the State Historical Museum in Moscow). Their correspondence is the first documentary evidence of a Slavic vernacular, since the older written language, Old Church Slavonic, was mainly a language of nobles and priests. But the letters are also precious because the oldest of them, dating to the eleventh century CE, were written before Balts and Finns had started to record their own languages, so they provide a glimpse of a city in whose streets literate and preliterate people mingled. We can be confident that the Balts were present, taking their cut of the lucrative sale of squirrel pelts, because the Baltic language Lithuanian borrowed a Slavic word for a now-obsolete unit of weight known in English as a 'ship-pound' (*birkavas* in Lithuanian, from Old Novgorod *bérkovec*).

The birch-bark letters speak to a highly literate population, since among the authors were women and children. Besides endless inventories of furs they include a marriage proposal, a dispute over a stolen female slave, and a little boy's drawing of himself as a brave warrior. The authors cursed, cracked jokes, spread scurrilous rumours and signed off with formulae such as 'I kiss you' or 'be so kind'. Happily for historical linguists, one fourteenth-century missive included a small Finnic–Slavic dictionary. The city in which the letters were composed was heavily fortified, with log-lined roads, and stone churches and monasteries that were treasure houses of icons and frescoes. Medieval Novgorod was known for its art, a reputation it maintained after coming under Muscovite rule in the fifteenth century. But it was that change of regime that ushered in a new East Slavic dialect, a forerunner of Modern Russian, and caused Old Novgorod to fall into disuse. At about the same time, Balts and Finns finally took up writing.

The Vikings, or Varangians as they were known in the east (probably from the Old Norse word *varing*, meaning 'ally'), would have spoken Old Norse to begin with, but like Viking invaders everywhere they took up the language of the people they conquered.[6] These formidable mercenaries in their clanking chainmail were among the chief enslavers of Slavs, yet they embraced the local Slavic tongue so quickly – within a couple of generations – that they barely had time to leave their mark in it. Their arrival in the east is nevertheless immortalised in the name 'Russia', since they were known in Novgorod as *Rus* (sometimes written *Rus'*, to indicate a soft *s* sound at the end). Linguists don't agree on where the name Rus came from, but one theory traces it to the Finns' name for Sweden, *Ruotsi*, whose literal meaning might have been 'oarsmen'.

In time the Varangians set up trading posts further south, notably at Smolensk and Kyiv. At the Dnieper Rapids they hauled their flat-bottomed boats out of the river and rolled them over logs until it was safe to put them back in the water. Influences flowed upstream too, from Constantinople, and in 988 CE a prince of Varangian heritage named Vladímir converted to Orthodox Christianity. Vladímir, which means 'conquer the world', is said to have had his entire court baptised in the Dnieper. The religion spread, and over the next few centuries Rus came to refer to all Orthodox Eastern Slavs. Only later did the name become attached to a geographical space, and later still was that space divided into Russia, Belarus and Ukraine. The Finns were Catholic, having been converted by the Swedes, but in Novgorod Eastern and Western Churches met, and it was here, from the Orthodox Slavs, that the Finns took their words for 'priest' (*pappi*) and 'cross' (*risti*). The Balts breathed the incense but rejected the creed. They remained stubbornly pagan, continuing to sacrifice horses, in rituals resembling the *ashvamedha*, right up until the thirteenth century. They even imported those horses, by boat across the Baltic, from their Christianised neighbours in Scandinavia.

NORTHERN IDYLL

The first Balts to convert were the Old Prussians, who had no say in the matter. The invasion of the crusading Teutonic Knights marked the beginning of the centuries-long decline of their language, as German-, Polish- and Lithuanian-speakers moved into their territory. The last speakers of Old Prussian probably died of famine and plague around 1700. The Latvians were next to embrace the Western Church, but the Lithuanians continued to hold out. After decades of vacillating between Eastern and Western options, Europe's last pagan nation capitulated in 1387, joining the Western Church.

It did so for reasons of political expediency. In order to marry the young Queen Jadwiga of Poland and claim the Polish throne, Jogaila, Lithuania's pagan Grand Duke, had to agree to convert himself and his country to Catholicism (he took the baptismal name Władysław, meaning something like 'famous ruler', and the dynasty he and Jadwiga founded became known as Jagiellon). The royal union ushered in four centuries of shared Polish–Lithuanian history, the last two of which saw the two countries form a single state. Balts and Slavs were united again, as they had been four millennia earlier, but not for long. Austria, Russia and Prussia divided the state up between them, and by 1800 the Polish–Lithuanian Commonwealth had ceased to exist.

The next time you find yourself in Kraków, Poland, go to the Wawel, the traditional seat of Polish kings, and head for the cathedral.† Just before you pass through the great door into the realm of dead royalty and some exceedingly fine stained glass, look up.

† *Wawel* is pronounced *Vavel* in English, while the girl's name *Wanda* is pronounced *Vanda*. As an ad campaign for a popular brand of Polish vodka made clear in the noughties, there's no *v* in *wodka*. In the Polish alphabet the letter *ł* is pronounced like English *w*, while the letter *w* is pronounced like English *v*.

Suspended by chains above your head is a jumble of monster bones. If they seem out of place, if they strike you as pagan, you are right on both counts. They are the bones of a whale, a mammoth and a rhinoceros that were pulled from the silt of the River Vistula (Wisła to the Poles) when the foundations of the cathedral were being laid in the eleventh century CE, though legend assigns them to a dragon called Smok.

A short walk across the Wawel complex, a hundred and thirty-five steps lead down in a spiral to Smok's supposed lair. You have to buy a ticket to enter, but if you ask nicely, the person who takes it from you will tell you the legend: how Smok terrorised the land, devouring livestock (or in some versions, maidens); how the sons of King Krak defeated the dragon, but then one son killed the other and was banished, so the kingdom fell to their wise and fair sister Wanda. Emerging from the cave, you find yourself on the bank of the Vistula, where a green-tinged statue of Smok breathes actual fire.

Smok or Żmij, or Zmei or Zmaj, depending on which Slavic-speaking country you happen to be in, is the archetypal serpent, denier-of-life, and any resemblance you may notice to J. R. R. Tolkien's dragon Smaug is not coincidental. Tolkien was a philologist who famously created *The Lord of the Rings* as a vehicle for languages that he had constructed, and he chose that name for its (to him) pleasing Indo-European ring. He knew that a natural language is always associated with a culture, and that every culture has its mythology. He therefore invented the legends that would make his languages live. (Tolkien referred to Esperanto and other constructed languages as 'dead, far deader than ancient unused languages, because their authors never invented any Esperanto legends'.[7]) There was actually a Proto-Indo-European word, **smeuk*, that probably meant 'to slide' or 'glide', and if the Slavic dragon names are derived from it then they are living exhibits of taboo deformation – the phenomenon whereby taboo words are rapidly recycled through euphemism and circumlocution. (In this case, a word describing the creature's motion may have come to replace its unspeakable name.)

During the communist period, Poles compared themselves to radishes: red on the outside, white on the inside – red and white being the colours of the Polish flag. The implication was that their communism was only skin-deep; beneath it they were Catholic, and capitalist. But beneath the Catholic layer is yet another: the pagan. The construction of the Wawel cathedral began just decades after Poland became officially Catholic, in 966 CE, and while it was being built a pagan rebellion got underway that nearly destroyed the nascent Polish state. Smok's bones serve as a reminder of that older layer, and of that near-annihilation.

In many Indo-European traditions the life-negating serpent became a milk thief, a 'cow-suckler' that wasn't past nestling in among the hungry calves, cuckoo-like, to claim its share. Biologically speaking this is impossible – snakes can't suck, and they lack the enzyme that would allow them to digest mammalian milk – but the idea is firmly cocooned in the Indo-European mind. A Hindi word for a reptile, *godhā*, literally means 'cow-suckler' – from *go*, 'cow', and *dha*, 'to suckle' – while Baltic names for a grass snake (Lithuanian *žaltys*, Latvian *zalktis*) are possibly derived from Proto-Indo-European ones meaning 'delighting in milk'. Here, again, the ancient nexus of Lithuanian and Sanskrit may reveal itself. But the Balts developed the cow-suckler motif in their own way, in their land of lakes and forests. They alone turned the snake into a friend, paying it in milk to protect the farm. If it wasn't fed it would unleash dark forces, but as long as you gave it what it wanted it would bring you luck. In pagan times, Baltic priestesses fed milk to snakes as part of their fertility rituals. 'It was a blessing to have a žaltys in one's home, under the bed or in some corner, or even in a place of honour at the table,' wrote Gimbutas in *The Balts*.

Perhaps it was inevitable, then, that the grass snake would become the symbol of pagan resistance to Christianity. Medieval chroniclers relate that the German crusaders seized the snakes from the village houses and threw them on to bonfires. After the Lithuanians converted in the fourteenth century, crosses began to spring up all over the Lithuanian countryside, but often they had snakes carved into them. Depending on your point of view, the snakes were Christianised, or the crosses were paganised. And although the Lithuanian language is far from the throwback it was once thought to be, the country's pagan past still lies close to the surface. Well into the twentieth century it was not uncommon to hear a Lithuanian say, *Kur žalčiai yra, tai ten tie namai yra česlyvi.* 'Wherever grass snakes are, the house is full of happiness.'

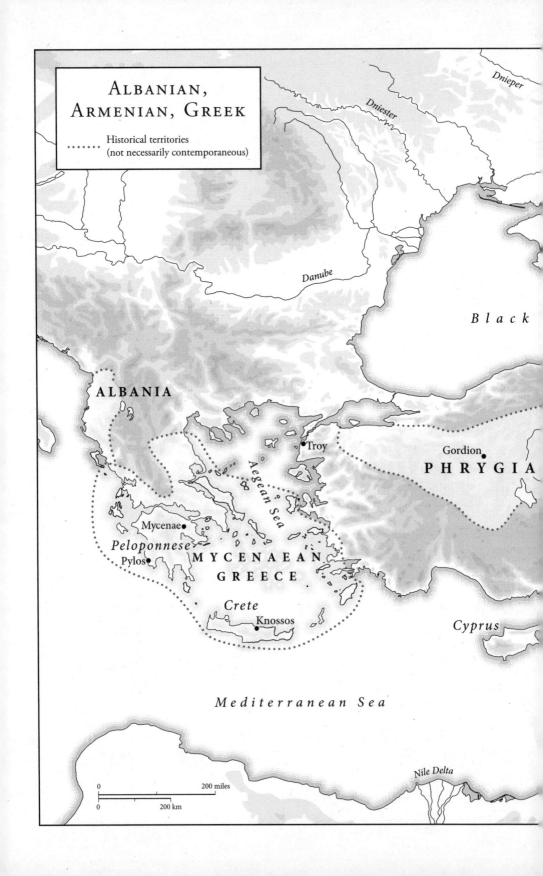

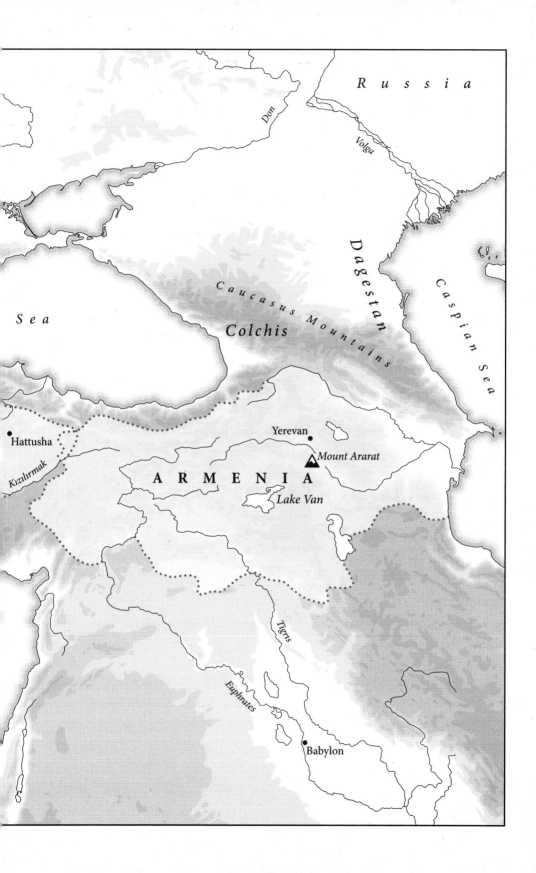

8

THEY CAME FROM STEEP WILUSA

Albanian, Armenian, Greek

And suddenly there they were, the long-haired Greeks, clad in bronze, stealing other people's women and brooding on eternal fame. Achilles, Helen, Agamemnon and the rest, the first protagonists of European literature, brought back to life by a poet or poets we call Homer. The Greek audiences who flocked to hear *The Iliad* took it for granted that those individuals had lived and breathed, and they were right, more or less. Homer may have embroidered their deeds and bloodlines, but he was singing about real people, contemporaries of the Hittite kings and the outsized personalities who inspired the Vedas. People who trod the shores of the Aegean hundreds of years before him, in the time of heroes.

Homer's inspiration, the models for Achilles and Agamemnon, belonged to the Mycenaean civilisation which arose around 1700 BCE and reached the peak of its splendour four hundred years later. These were the Greeks who came bearing gifts, the ones who won the Trojan War, but not long after that triumph they vanished from the historical scene – victims, it's thought, of the same wave of calamity that claimed the Hittites and brought the Bronze Age to a close. Greece was plunged into its Dark Ages, a period of chaos during which almost no art was created, but generations of preliterate poets nevertheless kept the stories alive in the way they always had, by word of mouth.

Eventually a new civilisation arose in Greece, one that had borrowed an alphabet from the speakers of a Semitic language. This alphabet would go on to inspire all modern European alphabets, a fact preserved in its English name, which fuses the first two Greek letters, alpha and beta. It was this same alphabet in which versions of the stories attributed to Homer were first recorded between the sixth and ninth centuries BCE. And it wasn't just the tales that were captured for posterity; *The Iliad* and *The Odyssey* are unique repositories of archaic expressions and poetic devices, threads connecting Homer's linguistic world to that older one. As well as being great works of art, his epics are key pillars of evidence that the real Achilles and Odysseus spoke an Indo-European language, an ancestor of Ancient and hence Modern Greek.

Besides their reputation, the Mycenaeans left many physical traces of themselves behind. Starting in the late nineteenth century with Heinrich Schliemann, an amateur who had earlier excavated at Troy, archaeologists have brought to light ruined palaces, the graves of heavily decorated soldiers and clay tablets that preserve scraps of the Mycenaean language in their own script. This was not the first script to be used in Greece. Before the Mycenaeans, the Minoans – the people named for the mythological King Minos, who inhabited the island of Crete – had used a script known as Linear A, which in turn grew out of older, hieroglyphic systems.[1] When the nearby volcano Santorini erupted around 1600 BCE, altering the climate and sending Minoan society into a tailspin, the Mycenaeans moved into Crete and developed their own script, modelling it on Linear A. The Mycenaean system, known as Linear B, was a syllabary, meaning that each of its symbols represented a syllable rather than a speech sound as in an alphabet. Linear B spread from Crete to the rest of the Greek-speaking world.

In 1936, a group of schoolchildren visited an exhibition in London marking fifty years of the archaeological research institute known as the British School at Athens. Sir Arthur Evans, the archaeologist who had excavated the Minoan Palace of Knossos several decades earlier, happened to be present. Then aged eighty-five, he proceeded to show the visitors certain finds from Knossos, including some Linear B tablets. 'Did you say the tablets haven't been deciphered, Sir?' one fourteen-year-old boy asked. His name was Michael Ventris, and ever since learning about the Egyptian hieroglyphs at the age of seven he had nurtured a passion for ancient scripts. Though he went on to become an architect and was never formally trained as a philologist, that passion, and his cryptographer's mind, led him to crack the Linear B code just sixteen years after meeting Evans.

Ventris might have hoped that the tablets contained the Mycenaeans' own version of events at Troy, but they turned out to consist mostly of receipts and IOUs. They are nevertheless among the oldest documents that exist in any Indo-European language, being about the same age as the *Rig Veda* and only a little younger than Hattushili I's will. Their immense value lies in the fact that they prove what Homer had hinted at: the Mycenaeans spoke an archaic form of Greek, not some unrelated language that had died out in the Dark Ages. Linear A has yet to be deciphered, so nobody knows what language the Minoans spoke, but the strong suspicion among linguists – bolstered in recent years by new kinds of evidence – is that it was non-Indo-European. The Minoans and the Mycenaeans were related to each other, as were their scripts, but the tongues they spoke were probably not. The language of Achilles came from far beyond the wine-dark sea.

How did Greek get to Greece? The language itself nudges us towards an answer. Greek is generally considered to belong to a 'Balkan' sub-group of Indo-European languages that also includes Albanian, Armenian and Phrygian.[2] It has more in common with these languages, in other words, and probably more shared history with them, than with other Indo-European languages. The exact relationships within the Balkan group are disputed, but the closest thing to a certainty is that Greek and Phrygian sprang from a common parent. Most historical linguists would also agree that Greek forms a kind of central hub in the group, with all the others being more closely related to it than they are to each other, and the Greek–Phrygian connection being the closest of all.

Connections beyond the Balkan group are more controversial. There are undeniable similarities between Greek and the Indo-Iranian languages. They modify verbs in the same way, to indicate that action took place in the past. There is striking overlap between their vocabularies too, especially when it comes to ritual and poetry. Their poets referred to the gods by the same shorthand, 'those who give riches', and certain of their deities had the same names modified only by the relevant sound laws (think Greek *Kérberos* and Indic *Śárvara*, mentioned in Chapter Six). But these similarities may only be apparent because Greek and Indo-Iranian boast such copious ancient literatures compared to other languages. And there are important differences between them too. Greek is a centum language that does not obey the ruki rule (according to which *s* becomes *sh* after *r*, *u*, *k* and *i*), whereas the Indo-Iranian languages are satemised (a hard *k* becomes a soft *s* before certain vowels) and do obey the ruki rule. Many linguists suspect that Greek and Indo-Iranian developed in the same neighbourhood, but that Greek and its Balkan relatives departed before satemisation and the ruki rule arose – after Tocharian, Italic, Celtic and possibly Germanic, but before Balto-Slavic and Indo-Iranian.

Eagle-eyed readers will have noticed that Armenian is included in the Balkan group, even though Armenia is nowhere near the Balkans. Despite the fact that this country lies south of the Caucasus and is surrounded by Turkic- and Caucasian-speaking populations today (with one other Indo-European language to the south, in Persian), its language's closest relatives are indeed to be found in the Balkans. This is why trying to plot the prehistoric itineraries of the Balkan languages is almost comically complicated. There are those who say that they were born in the Caucasus, like Anatolian. Armenian stayed put, according to this scenario, while the others struck out west across the Turkish peninsula to reach their ultimate destinations. Others claim that the entire group departed from the Pontic-Caspian steppe, heading anti-clockwise around the Black Sea. Albanian peeled off first, gravitating to its mountain realm, then Phrygian split from Greek and crossed the Bosporus to reach western Anatolia, from where an offshoot drifted east to root itself in the Armenian highlands, becoming Armenian. A third theory has the Balkan languages part ways on the steppe. Greek, Phrygian and Albanian headed west, while Armenian moved south through the Caucasus.

The fact that each Balkan language is alone in its branch doesn't help. Each of those branches was surely more crowded in the past, with phantom languages that died before they could be written down; but since there is no record of them, it's impossible to compare siblings and difficult to hazard a date for when each branch's proto- or parent language was spoken. On top of that, linguists have to deal with a very lopsided body of evidence when it comes to the Balkan group. Greek is very well attested, and from an early date, but mere fragments exist for Phrygian, and Armenian was only written down in the fifth century CE. Albanian was the last of all the known Indo-European languages to be documented, in 1462, by which time it had evolved so far from its roots that it takes a trained eye to see that it is Indo-European at all (linguists refer to it as the 'stepchild' of the Indo-European family for that reason). The Albanian words for 'six' and 'eight', *gjashtë* and *tetë*, don't look anything like their Latin counterparts *sex* and *octō*, even though they too descend from Proto-Indo-European *sék's and *ok'tó by predictable sound changes. We first 'see' the Balkan languages at very different stages of their evolution, in other words. Because of these difficulties, ancient DNA evidence has had an outsized impact on the Balkan story, upweighting certain scenarios with respect to others. If ancient languages tracked ancient migrations, then the hypothesis considered most plausible today is the third: the Balkan languages split on the plains north of the Black and Caspian Seas, and a predecessor of Armenian crossed the Caucasus. But by no means everybody agrees with that.

In the last few years, ancient DNA has given us our first glimpse of who Homer's heroes really were, and where they came from. The Mycenaeans were related to the Minoans in that both were descended from the first farmers to cross the Aegean from Anatolia, over eight thousand years ago. But they differed in one important respect: the Mycenaeans could also claim heritage in the steppe. They, alone of the two, were the products of interbreeding between farmers and nomadic migrants from the north. The migrants must have been the Yamnaya or their direct descendants, rather than the Corded Ware people who streamed into central and northern Europe, because Mycenaean males carried Y chromosomes that match those found in Yamnaya – but not Corded Ware – men. This is one of the reasons why scholars are now more confident that the Minoans spoke a non-Indo-European language: King Minos' subjects were not related to the steppe people who likely brought Greek to Greece.

As we saw in Chapter Five, the Corded Ware immigrants traversed Europe in a few generations and drove an almost complete replacement of the gene pool. They suppressed the indigenous population, likely by force, and kidnapped local women. The migration into Greece could hardly have been more different. It was gradual, probably not discernible within a single or even several generations. The steppe migrants blended thoroughly with the local farming populations, and there is no sense that they imposed themselves as an elite. In 2015, a tomb was opened at a site known as Nestor's Palace near the Mycenaean town of Pylos in the Peloponnese. It contained a man in his thirties who had been given a hero's burial, with riches worthy of Achilles himself. Among them was an ivory plaque carved with a griffin, a mythical creature that merged lion and eagle. The Griffin Warrior, as the individual is known, died around 1450 BCE, when the Mycenaean civilisation was approaching its zenith. He bore no steppe ancestry, but that ancestry has been found in some very humble Mycenaean graves. Whatever it was that granted you high status in Mycenaean society, then, it wasn't a blood link to the migrants from the north. Yet those migrants managed to impose their language. They are the reason that Greeks speak Indo-European.

The Yamnaya themselves didn't get as far as Greece. When they moved out of the Pontic steppe around five thousand years ago, heading west, most followed the Danube into what is now Hungary and regions north – as we also saw in Chapter Five – but some veered south into Serbia, Kosovo and Albania. (To archaeologists' surprise, Yamnaya burials have even been found in the mountainous parts of the Balkan peninsula. It's possible that early mixing with local populations allowed them to adapt quickly to that unfamiliar environment.) But they only ventured as far as the borders of modern Greece, so it must have been their descendants who carried their genes further south. Steppe ancestry first entered Greece around 2000 BCE, from the north-west. From there it spread, but without ever attaining the high proportions seen in central and northern Europe. By 1500 BCE it was present throughout the modern territory of Greece.

Greek is remarkable among Indo-European languages in that it has remained a single language for the more than three thousand years of its recorded existence (compare its contemporary Sanskrit, which engendered many daughters). It has long existed in different dialects, however, and by comparing these linguists infer that Proto-Greek was being spoken by 2000 BCE.[3] The date at which one southern Greek dialect, Mycenaean, stepped off the mainland into the islands is probably around 1500 BCE. The Minoans had abandoned Knossos a hundred or so years earlier, and for the first time the labyrinth where Theseus slew the Minotaur rang to Indo-European. Given that the Balkans received many small waves of immigration over time, it's possible that Albanian was seeded by a separate group of migrants from the one that seeded Greek, or that the two languages are different blends of older Indo-European dialects. The chariot reached Greece too, in this period, though again it might have been ushered in by people other than those who brought Greek (one suggestion is that it was a legacy of the Srubnaya, the culture responsible for the industrial-scale copper-mining in the Urals). What's clear is that two steppe innovations – an Indo-European language and the formidable Bronze Age war machine – collided in the Mycenaeans, and that the collision would shape European culture to this day.

Chariots appear throughout *The Iliad*, not only as lethal weapons but in races and, on one occasion, as the instrument of a shocking act of revenge. Having slain the Trojan prince Hector for killing his beloved Patroclus, Achilles slit Hector's ankles, threaded through some leather straps and hitched the dead man to his chariot. 'Then he lifted the famous armour into his car, got in himself, and with a touch of his whip started the horses, who flew off with a will. Dragged behind him, Hector raised a cloud of dust, his black locks streamed on either side, and dust fell thick upon his head, so comely once, which Zeus now let his enemies defile on his own native soil.'[4]

'Are you archaeologists?' I asked the American pair seated at the next table in the hotel restaurant at Boğazkale, who like me were delecting in a midday *meze*.[5] It seemed a good bet. The man had addressed the proprietor in fluent Turkish. We were in a small village whose only obvious attraction was the ruins of the Hittite capital, and, what can I say, they had an air of academia about them. 'Yes,' they replied, laughing, 'does it show?' A moment later I was chatting with Brian Rose, director of excavations at the Phrygian capital, Gordion, and his fellow archaeologist and Near Eastern expert Elspeth Dusinberre. They were taking a break from their work at Gordion to visit the capital of the older civilisation.

The Phrygians were those people once regarded as having spoken the first Indo-European language on the Turkish peninsula, until Bedřich Hrozný deciphered the clay tablets from Boğazkale in 1915 and showed that the Hittites had spoken one centuries earlier. They are thought to have come from the Balkans, crossing the Bosporus around 1200 BCE and reaching central Anatolia by the eighth century BCE. The Hittite Empire was no more, and they built their capital, Gordion, over the remains of a Hittite city. Gordion was named for its founder, Gordias, whose son Midas was immortalised in Greek mythology for turning everything he touched to gold. In a less well-known episode of his mythologised life, Midas was slapped with a pair of donkey ears for marking the god Apollo down in a music competition.

Recalling a documentary I had seen about the Phrygians, to which Rose had contributed, I asked him if he had found Midas yet. There is a huge burial mound at Gordion nicknamed the 'Midas Mound', that was excavated in the 1950s, but archaeologists later concluded that it was too old to be his. It was more likely to belong to his father, namesake of the fabled Gordian knot (the one that Alexander of Macedon sliced through in 333 BCE).

'Not yet,' Rose replied, so I scrolled through my phone to find a picture I had taken a few days earlier, of a skull on display at the Museum of Anatolian Civilizations in Ankara. The label read 'Midas' Skull'. Rose raised an eyebrow and told me that since the skull came from the Midas Mound, it was more likely to be that of Gordias. Then he told me that a minuscule fragment of that skull was about to be sent for genomic analysis. If any of Gordias' DNA had survived the more than three thousand years since he was laid in his tomb, the results might reveal something about his ancestry. A single individual can never give you the full picture, but I knew that archaeologists and geneticists were scouring Anatolia for other ancient human remains, and together with these Gordias could shed light on the prehistory, not only of the Phrygians and their language, but also of the Armenians and theirs.

Phrygian probably split from Greek in the Balkans, when the band of migrants whose descendants would found Gordion struck out for the Bosporus. The case for the Phrygians having seeded Armenian rests in part on Herodotus. 'The Armenians, who are Phrygian colonists, were armed in the Phrygian fashion . . .' wrote the father of history/father of lies. But although some linguists see significant overlap between Greek and Armenian, the linguistic evidence for a close relationship between Armenian and Phrygian is weak. It doesn't help that there is little to go on when it comes to Phrygian – mainly votive inscriptions and graffiti at Gordion, along with curses on anyone intending to desecrate tombs – and nothing at all for Armenian in the same period. Armenian wouldn't be written down until long after the Phrygians had vanished.

So far, studies in ancient DNA have only undermined the Armenian-from-the-west theory. As the steppe ancestry flowed down through the Balkans, it became diluted. There were ancient Albanians, Croatians, Serbians and Bulgarians who were practically indistinguishable from the Yamnaya, genetically speaking, but the Mycenaeans only carried about a third as much steppe ancestry. The Phrygians, apparently the first people to bring steppe ancestry into Anatolia (where, as we saw in Chapter Three, it had been all but absent throughout the Copper and Bronze Ages), had lower levels again. But prehistoric Armenians had very high levels of Yamnaya ancestry, around twice as much as the Mycenaeans, which is hard to explain if they were descended from Phrygians.

Meanwhile, the theory that speakers of a forerunner of Armenian carried the language across the Caucasus has been rising up the rankings. As early as the second or even third millennium BCE, Armenian and the languages of the unrelated Caucasian family loaned words to each other. Linguists can't always tell who lent what to whom, but they don't doubt the great age of these loans since they predated certain important sound changes in the languages concerned. (The direction of the loan of the 'wine' word – *gini* in Armenian, *ğvino* in Georgian – is particularly controversial, since both Georgians and Armenians claim to have invented wine. It is nevertheless possible that the word and the thing it describes have different origins.)

We also know that people were moving through the Caucasus, from the steppe, in the third millennium BCE. Geneticists have detected the southward flow of steppe ancestry from 2500 BCE, and they say that within five hundred years it had shown up in the Armenian highlands. It maintained an uninterrupted presence south of the Caucasus for the next thousand years or so, and the Y chromosomes that accompanied it persisted for even longer. Today, many Armenian men strolling past Soviet-era buildings in the capital, Yerevan, or drinking coffee in the city's many cafés, carry Y chromosomes that they inherited from the Yamnaya (the proportion of Greek men of whom this is true is much smaller, reflecting post-Bronze Age population displacements in Greece). Armenian men are, in the words of Iosif Lazaridis, a geneticist in David Reich's group at Harvard, 'literally the last male-line descendants of the Yamnaya people'.

The genetic transformation of the Caucasus between 2500 and 2000 BCE might have brought an Indo-European language into

Armenia. It certainly coincided with a major cultural shift in the region. For fifteen centuries prior to that, the societies that dominated much of the Caucasus were radically egalitarian. They seem to have recognised no difference in social worth between men and women, or between rich and poor. Adam Smith, an archaeologist at Cornell University, sees their collective burials, stripped of all status symbols, as a strong statement: a wholesale rejection of the hierarchical model of Uruk and the other Mesopotamian cities to the south. He suspects that the people buried in those communal graves may have been bound by a religion that enshrined egalitarian values.[6] But by 2500 BCE these prehistoric refuseniks had gone, and a very different culture had taken their place.

These new societies were radically *un*equal, but not in the way that Uruk was. They were societies of cattle-herders and horse-breeders who buried their dead individually, beneath kurgans. Women and men received different treatment in death, and some of the kurgans were, in Smith's words, 'unbelievably garish'. The largest would have taken hundreds of person-hours to build, and they were filled with precious metals and sacrifices, including human sacrifices – something the region hadn't known until then. The burials speak to an uptick in violence too, often in the form of sword wounds.

If this was the moment when the Indo-Europeans arrived in the Caucasus, it could have been they who left the dozens of basalt stelae that litter the Armenian highlands at altitudes of over two thousand metres (between six and seven thousand feet). These 'dragonstones', or *vishaps* as they are known locally, have now mostly toppled on to their sides, but they once stood up to five metres (sixteen feet) tall. Many were carved to look as if the hide of a horned animal had been draped over them, prompting some to draw parallels with the animal sacrifices that the Yamnaya once performed in the flatlands to the north.

By 2000 BCE the Yamnaya had gone from the Pontic-Caspian

steppe, and their descendants had been replaced there by those of the Corded Ware culture or the Sintashta further east. It's possible, though by no means proven, that the future Armenian-speakers, the last males in the Yamnaya line, departed with their clans just as this reconfiguration was taking place. Leaving behind neighbouring populations of Proto-Greek-speakers to the west, and Proto-Indo-Iranian-speakers to the east, they headed for the mountains, hugging the Caspian coast of what is now Dagestan to bypass the higher peaks. As time went on they interbred with the people they encountered, who spoke Caucasian languages, and little by little their beliefs and their burial rites changed. But they retained the elements of ritual and poetry that they had once shared with the future Greeks and Aryans.

Detecting this migration is one thing. Connecting it to the birth of the Armenian language is another. If you're standing in the present and looking back, it's not easy to discern the moment at which the Armenians first appear. The name 'Armina' is preserved in Persian and Greek inscriptions from the sixth century BCE, where it refers to a Persian satrapy or vassal state. The Persians had just defeated the Urartians, who spoke a non-Indo-European language related to Hurrian, and who had controlled the Armenian highlands for several hundred years before that. One theory holds that the Urartians had ruled over an indigenous, Armenian-speaking population, and that the Armenians seized the opportunity of the Persian conquest to throw off the Urartian yoke and be recognised in their own right. (Eventually, they would declare their independence from Persia too.)

But Armenians don't call themselves Armenians, or their country Armenia. They are *Hayer* from *Hayastan*, and traditionally they trace themselves back to a tribal confederation that entered history long before the 'Arminans' mentioned by the Persians. This was the Hayasa, who lived close to Lake Van and fought the Hittites in the thirteen hundreds BCE. (Lake Van is in eastern Türkiye today, but along with Mount Ararat it was once part of Armenia.) We know about this conflict because the Hittites wrote about it. Unfortunately the Hittites didn't tell us anything about the Hayasan language, but if the Hayasa spoke a predecessor of Armenian, they were there long before the Phrygians arrived – further undermining the theory that Armenian came from the west.

There is one last, slender strand of evidence to add to the other, more robust ones that Armenian came through the Caucasus. The mythical founder of Hayastan was a handsome, friendly giant called Hayk. You can see a statue of him today, in Yerevan, extending his mighty bow. Hayk is supposed to have stood up to the Babylonian tyrant Bel and to have led his large family into exile. Setting off 'with his sons and daughters and sons' sons, martial men about three hundred in number', he stopped first at the foot of Ararat.[7] Then, leaving that region to his grandson, he moved on towards Lake Van and settled to the north-west of it. When Bel tried again to get Hayk to submit to him, offering him the command of his youthful warband, Hayk again refused and called Bel 'dog'. A battle ensued in which Hayk killed Bel, and at the site of his victory he built an estate: Armenia, or Hayastan.

Armenia was the first country to adopt Christianity as its state religion, in 301 CE, and Hayk's story wasn't written down until after that. The earliest preserved version of the myth dates to the fifth century and already carries a biblical gloss. Hayk is presented as a descendant of Noah, and his nemesis, Bel, is equated to Nimrod, who the Bible says built the Tower of Babel (indeed, Armenians celebrate Hayk as the only one to oppose the plan to build that tower). But some scholars have noted that beneath the gloss the story harbours some very Indo-European themes, from the patriarch at the head of his clan to the canine slur and Bel's band of hunting youths. And although the myth tells us that Hayk led his family away from Babylon, its internal geography contradicts that. Hayk reached Lake Van by way of Ararat, as if he had been travelling south-west. Extrapolate that path backwards and one might be tempted to think that he was coming, not from Babylon in the south, but from the mountains in the north.

Albanian, like Basque, has to some extent been protected by its mountain situation. It likely once had relatives, in its branch, that vanished during the many reconfigurations of Europe's linguistic landscape over the last five thousand years. Albanian, alone of those siblings, resisted. It withstood the Roman Empire, the arrival of the Goths (who left it a word for 'trousers', *tirq*) and that of the Slavs who followed the Goths south. If you look at a linguistic map of the Balkans today, you'll see that Albanian is surrounded by a sea of Slavic languages – Bulgarian and Serbo-Croatian among them – except at its southern frontier where it comes up against Greek.

To say that Albanian *resisted* these onslaughts is true, much as English resisted Romance, but also in some ways misleading. History certainly left its mark in Albanian. It borrowed vocabulary from Ancient Greek, Latin, Slavic, Turkish, Romani and Italian, as well as from its direct neighbours in the Balkans. It has been estimated that, today, originally Albanian words comprise only ten per cent of the Albanian lexicon.[8] The Latin loans entered after the third century BCE, when the Romans crossed the Adriatic and pushed east, but so much time elapsed between their absorption and the earliest recording of Albanian that linguists failed to recognise their Latin-ness until the nineteenth century. The Albanian word *mik* (friend), they now agree, comes from Latin *amicus*, while *mbrët* (king) is actually Latin *imperator* – which referred to a high-ranking soldier before it referred to an emperor – transformed by predictable sound changes over two thousand years.

Albanian has slowly been converging with its Balkan neighbours in its grammar and sound, matching what has happened between Basque and its Romance neighbours. In the early centuries of the Common Era Basque was influential in forming the Romance languages of south-west Europe – especially Spanish and Occitan. These days, however, the influence is almost exclusively in the other direction. Lexically, Basque is now considered a mixed language, since over half its words are of Latin or Romance origin. Are Albanian and Basque fighting a losing battle, or adapting to survive? Perhaps the best way to describe them, though their situations differ in important ways, is as holding up remarkably well two thousand years, or more, into an unrelenting siege.

> And once again Albania cowered in a hut
> In her dark mythological nights
> And on the strings of a lute strove to express something
> Of her incomprehensible soul,
> Of the inner voices
> That echoed mutely from the depths of the epic earth.[9]

Archaeology has long provided a counterweight to the testimony of writers from the classical period, but the ability to read ancient DNA, and to reconstruct extinct languages, has put those writers to the test like never before. In general, they are turning out to be reliable guides – sometimes more reliable than they have been given credit for.

Take Homer. He sings of a world united behind a single ruler, Agamemnon, 'king of men'. Rival factions may have strained at the leash, but they did their leader's bidding in the end. They went to war when he asked them to. In this aspect of his portrayal of the Achaeans – Homer's name for the Mycenaeans – scholars have long suspected him of poetic licence. They consider that Bronze

Age Greece had a number of power centres, among them Pylos and Agamemnon's fief Mycenae, that were too small and inward-looking to pose any serious threat to the major powers of the day. Homer, who certainly lived in a world of squabbling city-states, must therefore have been looking back at the past through rose-tinted glasses. Over the last few decades, however, an alternative view has been gaining ground: that Homer was right, and Mycenaean Greece *was* a state worthy of being mentioned in the same breath as the contemporary empires of the Near East – the Hittites, Babylonians, Assyrians and Egyptians.

This subversive proposal is inspired in part by an absence of evidence. The documents that have been preserved from Bronze Age Greece conjure local bureaucrats who saw no further than the palace they served and the few surrounding villages that kept it in olive oil. Their instincts were tribal and their 'cities' glorified farmsteads. These 'Potemkin palaces', as one scholar disdainfully dubbed them, hadn't moved far beyond the state of chiefdoms.[10] In a confrontation with the mighty Shuppililiuma, or Ramses the Great, they would have been swatted like flies.

Already in the 1950s, however, Michael Ventris had warned against reading too much into those documents, or rather too little. They were all written on clay, but clay was not, in his opinion, the ideal medium for the fine lines and delicate curves of Linear B. He suggested that it may have been used for rough drafts before the texts were transferred to papyrus or animal skin, on which those texts were reproduced with pen and ink. Since the paper or skin had not survived, and since clay tablets could be recycled, there might once have been a much larger body of documents than the one that had come down to us. More of the population might have been literate than archaeologists had allowed and, crucially, the palaces might have been in communication with each other.

A Dutch scholar, Willemijn Waal, has pushed Ventris' thinking one step further. The Mycenaean scribes shaped some of their clay tablets like palm leaves, and she thinks this may be a clue that the texts were ultimately transferred to real palm leaves. The Greek term *phoinikeia grammata*, usually translated as 'Phoenician letters' and assumed to refer to the Semitic alphabet that the Greeks adopted in Homer's day, could in fact have referred to the palm leaves that they had written on for centuries before that, since *phoinix* can mean both a palm tree and the land of Phoenicia (roughly modern Lebanon, where Semitic languages were spoken).[11] It's a controversial theory, and difficult to test since no leaf letters survive, but the implication is that the palaces were regional hubs in a larger state – one, perhaps, led by a single ruler.

This novel way of thinking about Bronze Age Greece doesn't rest only on the materials that scribes wrote upon. Jorrit Kelder, an archaeologist and colleague of Waal's, points out that some Mycenaean palaces contain paired structures that he interprets as throne rooms – one for the local governor, one for the visiting supremo. Hittite texts refer to the 'Great King' of a foreign country that they call *Ahhiyawa*. Very few leaders were granted the title Great King in the Near East at that time, because it signified real power, including the power to declare war on, and sign treaties with, foreign states. Ahhiyawa is now generally accepted to be the same as Homer's *Achaea*. The first Ahhiyawan the Hittite scribes mention is *Attarissija*, which looks suspiciously like Atreus or his sons the Atreids, Agamemnon and Menelaus. The scribes also discuss a city called *Wilusa* that was probably Troy – the city the Greeks knew as *Wilios* (the *W* was eventually dropped giving *Ilios*, after which *The Iliad* is named), and that both Anatolian and Greek poets described as 'steep'.

We know that the Mycenaeans ventured into western Anatolia, where Troy lies. Traces of them have been detected in the region as early as the fourteen hundreds BCE. If the Ahhiyawans really were the Greeks, and if Wilusa really was Troy, then the Hittite archives record that Hittites and Greeks fought several wars over that city in the next two centuries. Any of those wars, or all of them, could have inspired *The Iliad*. The poet could have left us a portrait of Greece in the time of Agamemnon that was close to reality – in which case, once that Greece vanished, its unity and sophistication weren't retrieved for a very long time.

What of Herodotus; should we also give him the benefit of the doubt? He was a historian, not a poet, so he doesn't have the excuse of poetic licence. He still told what we would consider some pretty tall tales – describing one tribe as sporting a single eye in the middle of their forehead, and another as uniformly bald. But Herodotus wasn't a historian in the modern sense. He saw it as his duty to convey the sources to the reader, not to select from them or to decide which was right. Coming from another era and another understanding of history, we have to learn how to read him. Even if his descriptions of those tribes were fanciful, they may have existed. And if they existed, then we have to admit that we know very little about many of them. We are a long way from understanding the multiplicity – including the linguistic multiplicity – of the ancient world. The work goes on, then, with the pleasing prospect that in decades to come, modern scientists will work hand in hand with the chroniclers of the classical age.

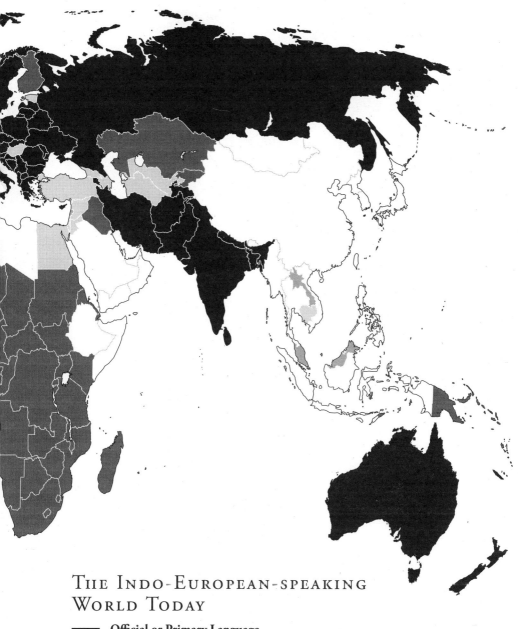

The Indo-European-speaking World Today

Official or Primary Language
Countries where Indo-European is a primary de facto national or official language

Secondary Official Language
Countries where Indo-European is a secondary official language

Recognised
Countries where Indo-European is officially recognised

Significant
Countries where an Indo-European language has a significant number of speakers or has a significant role

CONCLUSION

Shibboleth

In the summer of 1930, as construction was about to begin on the mount for the blast furnace at the future Azovstal metalworks in Mariupol, Soviet apparatchiks discovered what looked like an ancient burial there. They called in an archaeologist, Mykola Makarenko, and once he had confirmed that the burial was indeed of rare antiquity they asked him to carry out a rescue excavation.

Makarenko quickly realised that he was dealing with a large communal grave dating back to the late Stone Age. Burial after burial came to light, dozens of them, laid out in neat rows, their heads facing alternately east and west. Among the rich grave goods that had been buried with them, including boars' tusks and beads of porphyry and mother-of-pearl, his practised eye detected influences from Siberia and the Caucasus. As his team toiled, the work on the plant went on around them. 'Every day we were requested to hasten our excavation works,' he wrote. 'Crowds of workers surrounded the place of our work. They arrived here by hundreds from different factories and were interested in what they saw of our excavations. We were finally obliged to build a fence of barbed wire in order to prevent them from hindering us in our work, and to fix special hours for their inspection of the excavated material under the guidance of our expedition members.'

Besides the communal grave they discovered a younger burial of three individuals. It looked as if some later prehistoric people had returned, respectfully, to bury their dead in the place of their ancestors. Makarenko reasoned that they might also have been attracted by the site's natural defences. It was on a peninsula that was bound on two sides by the River Kalmius, on the third by the Sea of Azov and on the fourth by a ravine. In mid-October, after two months of intense activity, the archaeologists packed up their tools and left. They took with them one of the older graves and the younger, triple grave. These they deposited at Mariupol's museum of local history. The rest of the burial ground was razed to make way for the plant.

In 1938, for opposing the demolition of a medieval monastery in Kyiv, Makarenko was executed by the NKVD, the Soviet secret police. Seven decades passed, and the graves at the Mariupol museum were forgotten. Science advanced. Techniques were developed for extracting and reading ancient DNA. In 2010, Ukrainian archaeologists rediscovered the graves, and word of their existence reached the Ukrainian-born geneticist Alexey Nikitin in the United States. It took him another decade to persuade the museum's curators to part with some of the skeletal material, but they eventually agreed to let him have a few bones and teeth for analysis.

That was in November 2021. The following February, Russian forces invaded Ukraine and laid siege to Mariupol. Ukrainian troops defending the city retreated to the Azovstal plant with its man-made defences in the form of tunnels and bunkers, and its natural defences in the form of a river, a ravine and the Azov Sea. Thousands died in the fighting, some of whom were buried in a mass grave at the edge of the city. By the time the survivors surrendered in May, the metalworks had been almost obliterated, as had the history museum. All that remained of the ancient burial site, besides Makarenko's detailed notes, were the bones and teeth in Nikitin's custody.

CONCLUSION

His analyses of the chemical composition of those bones and teeth confirmed Makarenko's intuition that far-reaching interactions were at play in the Mariupol cemetery. Some of those people had travelled long distances in their lives. The sequence of burials had, Nikitin suspected, captured a key juncture in the genetic formation of the Yamnaya, the moment when a mix of Ukrainian and Siberian ancestries received an infusion of genes from the Caucasus. Only genomic analysis could confirm that, but Nikitin couldn't extract any usable DNA from the material, and there was no hope of getting more samples from Mariupol now. The prehistoric graveyard had gone, and its secrets with it.

It would be an understatement to say that Russia's war on Ukraine has had a detrimental effect on research into the Indo-European languages, whose cradle many scholars believe lies between the two countries. Landmines now infest Ukrainian soil like lethal spores, placing archaeological sites off-limits if they haven't already been destroyed. Some Ukrainian archaeologists have gone to fight, others have attempted to protect their cultural heritage. Russian scholars, even those who have spoken out against the war, have been frozen out of international collaborations and conferences, and no more samples have come out of the war zone since hostilities began. After a few decades of openness and collaboration between old Cold War enemies, channels of communication have frozen again.

And yet, despite the many practical and political obstacles, despite tensions over the ownership of data and the authorship of papers, new information continues to be published, based on materials that were collected before the war. For many the mission seems more urgent than ever, given that this war is, in part, a war over language – over where the russophone sphere begins and ends. Historical linguists carry on their work more or less unimpeded, piecing together the past, not only of Indo-European, but also of the other great language dynasties with which it has crossed paths.

The stories of Uralic, Sino-Tibetan and Afro-Asiatic, to name just three, are turning out to be just as complex and fascinating.

There are many questions still to answer, not least, what caused a band of brothers to head out from their ancestral valley on the Pontic-Caspian steppe and invent a new way of life – one that would propel their genes and dialects throughout the Old World? The evidence that might help scholars answer that question has eluded them so far, but it's not impossible that the war could bring it to light. A site called Mykhailivka, source of the oldest Yamnaya samples to date, was partially flooded when the Kakhovka Dam was completed on the Dnieper in 1956. The destruction of the dam in June 2023 unleashed a human and environmental disaster, but Mykhailivka is reported to be dry once again.

Some things have been established beyond much doubt. The people who spoke the parent of all modern Indo-European languages carried a blend of ancestries that came together within an international trade network. They were nomads, migrants, who spread that language wherever they went. There was nothing inherently successful about their language. If they managed to impose it on the populations they encountered, it was because it enshrined a suite of values, myths and conventions that allowed them to expand and adapt. They thrived, and others wished to emulate them. Many of the offspring of that first Indo-European language died. The ones that survived were those whose speakers proved adaptable in their turn. They did not stay the same, nor did their languages. That was the secret of their success.

My husband's name is Richard, though he sometimes writes it the Polish way: Ryszard. His parents were forced to leave Poland in the dark days of the Second World War, and for many years they expected to return. They met, married and raised their children in

London. Richard grew up speaking Polish at home and English outside it. The first time that he and I went to Poland together, I remember the curious looks that darted his way. Finally a woman said what, presumably, others were thinking: 'You speak without an accent but there's something odd about your Polish, I can't put my finger on it.' Richard told her what it was: their two versions of the Polish language had parted ways in 1940. His branded him the son of émigrés.

The human compulsion to communicate is overwhelming. It was there even before *sapiens*. Children born deaf have invented sign languages to escape what one researcher has called the 'abysmal loneliness' of life without communication.[1] Some argue that without language there is no reasoning, others that there is no consciousness. But besides the desire to reach out to strangers, there's another one that has deep roots in us: the desire to belong to our own group.

People change languages just by using them. Sometimes they do so unconsciously, by reproducing a sound that they have heard infinitesimally altered. Sometimes they do so consciously, to imitate or differentiate themselves from others. The accumulation of these conscious and unconscious changes causes languages to split. There is some evidence that languages change most rapidly just *after* they split, as the two groups assert their diverging identities. From then on, the moment a newcomer opens their mouth, they reveal whether they belong to 'us' or 'them'.

In Belfast, a student called Kacey told me that when young people in Ballymena, County Antrim, say they're 'skundered', they mean they're confused. Elsewhere in Northern Ireland, to be skundered is to be embarrassed. She didn't know how or why the difference had come about, but it clearly set the youth of Ballymena apart. Most of the time such differences don't matter. We all have a number of identities. A person can feel both Basque and Spanish, both Kurdish and Iranian, from Ballymena and from (Northern) Ireland, from London and from Poland. The identity that comes to the fore depends on the situation, but it's not something to go to war over. Until it is.

In the Bible, Judges 12:5–6, the men of Gilead identified their Ephraimite enemies by making them utter a single Hebrew word meaning 'stream': 'Say now Shibboleth: and he said Sibboleth: for he could not frame to pronounce it right. Then they took him, and slew him at the passages of Jordan.' Linguist Laada Bilaniuk reports that since Russia invaded Ukraine in February 2022, Ukrainian troops have been using their own shibboleth to identify undercover Russian operatives. The word is *palianytsia*, which means a loaf of bread. It may also have served as a shibboleth in the First and Second World Wars, Bilaniuk says. The difference in pronunciation didn't go away in between times, but Ukrainians and Russians didn't pay attention to it. The similarities between their languages, which are great, mattered more.[2]

Language shapes identity and identity shapes language. But to get at group membership in prehistory scholars have to rely on pots and bones, and admit that the finer distinctions escape them. They can be sure that prehistoric people yearned to belong, and vetted strangers as we do, but they can't know exactly what notion of 'us' they were guarding. In his book *Who We Are And How We Got Here*, David Reich describes a contretemps he had over just this issue. It was 2008, and having identified an influx of steppe ancestry to India over three thousand years ago, he and his Indian co-workers, Lalji Singh and Kumarasamy Thangaraj, disagreed about what to call the people who had brought it. Reich wanted the label 'West Eurasians', but Singh and Thangaraj pointed out that there was no way of knowing that the immigrants themselves had actually come all the way from West Eurasia. Though they certainly carried West Eurasian ancestry, and this set them apart from the existing population of India, they might have regarded themselves as Bactrians, or Gandharans (two ancient civilisations of Central and South Asia), or even Harappans. The only certainty was that they had once been *there*, where their remains were found. Reich conceded the point, and the group agreed on the label 'Ancestral North Indians' (as opposed to India's existing population, 'Ancestral South Indians').

This question of identity matters, because if scientists can't know how prehistoric people saw themselves, neither can anyone else. Yet there are plenty who would claim to do so for political purposes. The best preserved site of the Sintashta culture – the culture that many scholars associate with the Proto-Indo-Iranian language – is called Arkaim. It has been declared a national monument in Russia, where it is sometimes referred to as the Aryan homeland. Arkaim is central to the revival of the notion, dear to some in that country, of a Russian world (*Russkiy mir*) from which sprang all the living eastern branches of the Indo-European language family – Baltic and Slavic, even Indic and Iranic. (In the early 1990s, soon after Arkaim was excavated, you could see a poster at the entrance to the site that read 'Zarathustra was born here'.) There's no doubt that Russian president Vladimir Putin subscribes to this view. In 2021 he published an essay in which he stated explicitly that his goal was to reunite a once supposedly russophone sphere. But it's not just Putin's mistake. Nationalists across the globe are projecting national identities back on to a world that didn't know them. Prehistoric people undoubtedly had identities as complex and multi-layered as ours. We don't know what they were, but we can be sure that nowhere among the layers was the nation-state.

As I was beginning the research for this book, I stumbled on the work of a French artist called Gabriel Léger. He had recently completed a project that involved repolishing antique bronze mirrors from Greece, Rome and the Etruscan heartland, so that you could see your face in a surface that, twenty-five centuries ago, had reflected someone else's. The idea stayed with me. In fact I was reminded of it constantly as I travelled around (you'd be surprised, or perhaps you wouldn't, by how many tarnished mirrors there are in the archaeological museums of the world). We know that ancient people looked at themselves in mirrors. We don't know what they saw.

By the seventh century of the Common Era, Indo-Iranian languages dominated a strip of the Earth that swept from the southern Caucasus to Assam at the north-eastern tip of India. The Turkish peninsula was still home to Greek, Anatolian and Armenian, but the broad delineations of Europe's linguistic landscape looked much as they do today: Romance in the south and west, Germanic in the north and centre, Celtic on the edge. Part of the centre and much of the east were dominated by the Slavic tongues, the Baltic languages were shrinking towards the eponymous sea, and somewhere in the Babel of the Balkans one could make out Albanian and Greek.

The linguistic frontiers would continue to shift, and there was plenty of change to come within them. The imminent Arab conquests – which extended both east and west of the Arabian peninsula – would have varying success in supplanting the languages of the conquered peoples, like the Roman invasions before them and the Mongol invasions after them, but they would leave none of them unaltered. From the eleventh century Anatolia would receive an infusion of Seljuk Turks, whose ancestors hailed from the Altai Mountains, and they would gradually impose their Turkic tongue. The Jewish diaspora would create blends of Hebrew and Indo-European, of which the best known are Yiddish (a Germanic language) and Ladino (a Romance language derived from Old Spanish). And in the twelfth century, a thousand years after they left India, the Roma finally reached Europe.

What is striking about these linguistic encounters is that, as far as the data allow us to tell, the outcome was different in each case.

CONCLUSION

It's hardly surprising that the linguistic impact of immigration depends on the relative size of the host and incoming populations, how long the immigrants stay and whether they come in peace or violence. But linguists know that many other factors play a role too, including immunity to infectious disease; religion; house prices; job openings; education; roads; and much, much more.

'Once it is realised that the direction in which languages change is at the mercy of the arbitrary events that shake the lives of their speakers, the futility of searching for natural or even universal tendencies in language change becomes evident,' linguist Peter Schrijver has observed. 'This is a sobering thought.' Indeed it is. But it might not hold for very much longer. At the moment, historical linguists build their family trees on the basis of shared vocabulary. They would like to be able to integrate other components of language – notably sound and grammar – to understand how language changes as an *ensemble* over time, but the sheer volume of information involved has precluded that until now. Artificial intelligence may be a double-edged sword, but it excels at finding patterns in large bodies of data (what brilliant polyglots like Dante did for us, when the data were fewer). It will make that goal achievable, and then it will make it possible to explore in detail how the circumstances in which human groups meet affect the languages that their grandchildren will speak. Linguists will begin to discern rules governing language change, which will become increasingly predictable.

Until then, there will continue to be disagreement over whether all languages are born equal. Take the creoles, of which several dozen are spoken around the world, including French-based Haitian Creole, English-based Jamaican Patois and Goa Portuguese Creole, which as its name suggests is based on Portuguese. Many creoles emerged in the seventeenth and eighteenth centuries when Europeans exported their languages to their colonies around the Atlantic and Indian Oceans. They have traditionally been regarded as slightly less than full languages, simplified and then recomplexified versions of the ones spoken by the predominantly white settlers: not exactly *new*. But some argue that this view comes loaded with nineteenth-century prejudice, and that creolisation is simply language genesis by another name.

The transformations that certain Romance and Germanic languages underwent on colonial plantations are not so different from the one that German underwent when it came to England in the fifth century, according to this view. British Celts, enslaved or at least relegated to second-class citizenship by the incoming Anglo-Saxons, tried to speak German and produced English instead. French was what came out of the mouths of Germanic-speaking Franks when they took up Latin, and not classical Latin but the vulgar form already bent out of shape by the Gauls. Aside from the fact that each arose in a specific historical context, and that creoles were rarely written down, some see no difference.

Starting in the 1980s, one imperial language began to nose ahead of the field, and then to lap the others. English has been, so far, the sole beneficiary of the new era of globalisation, the first truly global language. Some have gone so far as to brand it a 'killer', on the grounds that it has driven many smaller languages to extinction, but that is not a label that sits easily with everyone.[3] Salikoko Mufwene, a Congo-born linguist at the University of Chicago, points out that English has expanded mainly as a lingua franca. It may have squeezed other lingua francas, such as Swahili in Africa or Malay in Asia, but it hasn't dented the indigenous languages that are spoken day to day in those places. The 'killer' label reflects a very Eurocentric outlook, Mufwene says, because it is Europe that has made a speciality of monolingualism. In much of the rest of the world, stable bilingualism or even multilingualism is still the norm.

Besides, English is simultaneously diverging into varieties that may one day be unrecognisable as the same language. So is English killing, or is it dying, or is it somehow doing both at once? It is true that there are more and more varieties of English, says Australian linguist Nicholas Evans, but it's unlikely to go the way of Latin, which exploded in a starburst we call Romance. Few in the late Roman Empire could read or write, and ordinary people living at opposite ends of that empire were not in direct contact with each other. The centrifugal forces pulling their vernaculars apart had no centripetal force to counteract them. Thanks to television, the internet and social media, the speakers of different English vernaculars are exposed to each other's speech, and to a standard written form of the language. Though it's difficult to predict a language's future, Evans says, English could well settle into a state of *diglossia*, where a gulf exists between the shared written form and the many spoken varieties, but the two bind each other together into a single tongue.

These debates matter, in part, because they have a bearing on the question of language erosion. Of the roughly seven thousand languages that are spoken in the world today, nearly half are considered endangered. Some assessments suggest that fifteen hundred could perish by the end of this century, and attempts at resuscitating the dying ones have proved, by certain measures, disappointing. The shining exception is Hebrew, which was reclaimed in unique circumstances (the birth of the modern state of Israel). But despite decades of intensive language-teaching and strong grassroots support, fewer than twenty per cent of Welsh people spoke Welsh in 2022, and the number had shrunk over the previous decade. Welsh is considered the healthiest of the surviving Celtic languages. The Irish that is being reinvigorated in Ireland is, experts say, heavily influenced by English.

> so many languages have fallen
> off of the edge of the world
> into the dragon's mouth . . .[4]

Not everybody finds these trends discouraging. Some point out that in prehistory, a language at the peak of its health had as few as a thousand speakers. Revival shouldn't be measured by number of recruits, therefore, but by the extent to which the language is the vehicle of a vibrant culture, and whether young people are taking it up. It's inevitable, too, that a language that has been reclaimed or reinvigorated should differ from its original form; that's evolution. (The Israeli-born linguist Ghil'ad Zuckermann argues that instead of Hebrew being revived, a new, Hebrew-European hybrid language was born. He calls it Israeli.)

Keeping endangered languages alive seems an indisputably good thing, if it's what their speakers want, and language activists are getting better at it. They've understood that the solution isn't simply to plough ever more resources into teaching. First they have to work out why people are abandoning the languages, then they have to address the inequalities that are causing them to do so. Language is a tool: it lives for as long as it's useful, for as long as it opens doors for its speakers and equips them to improve their lives. And while nobody denies that language deaths outnumber language births at present, it's important to take any numbers attached to the erosion phenomenon with a pinch of salt. Because language is so entwined with identity, any tally is at least partly subjective. Languages are changing on their speakers' lips as I write. Since most of those changes go unrecorded and unstudied, and linguists can't agree on what constitutes language genesis, the deaths may have been exaggerated at the expense of the births. It's time we took note of those hubs which, like hot vents at the bottom of the sea, have been churning out new linguistic life for decades now, and asked: are we on the brink of a linguistic renaissance?

CONCLUSION

Tacked to the wall of the car hire office in Varna is a map. Between the city of Burgas to the south and the Turkish border, an area has been hatched red. The accompanying text warns clients: don't pick up hitchhikers here. Bulgaria's border with Türkiye, most of which is now protected by a metal fence, also marks the European Union's south-eastern frontier. Only a hundred kilometres (sixty miles) beyond it lies the Bosporus, bridgehead for intercontinental migrants since the Stone Age. For four hundred years up until the Second World War, in the colonial period, the net flow of people was away from Europe. Since that war, it has been towards Europe. Today the migrants come predominantly from Syria, Palestine, Iran and Iraq; from Afghanistan and Pakistan; from North and West Africa.

I leave the office, my business concluded, and head out on foot into the falling dusk. It's November 2022. A good number of Ukrainian numberplates are interspersed with the Bulgarian ones in Varna's rush-hour traffic. Draped across the façade of the decommissioned Hotel Odessus, just behind the sea garden, is a blue-and-yellow banner that demands peace for Ukraine. The Golden Sands beach resort, a little way up the coast, is providing shelter for thousands of Ukrainian refugees. They have been protesting the government's decision to rehouse them. All they want, they say, is stability for themselves and their children.

The Armenian capital has absorbed its fair share of foreigners by the time I reach it the following June. The people seeking refuge there are mainly Russians, fleeing conscription or sanctions or simply a war that they don't condone. Historically, Russians have been welcomed in Yerevan, but the welcome will cool in the months to come as Russia fails to back Armenia in its long-running conflict with Azerbaijan. It's a different story in the Georgian capital. The Georgian government sees Russia as its natural ally, but the population longs to be closer to Europe, and graffiti covering Tbilisi expresses uniform hostility towards the northern neighbour. Leaving a café one day, I notice the following words printed at the bottom of my bill: 'Fact check: twenty per cent of Georgian territory is occupied by Russia.'

In July 2023 the mercury is climbing on the southern Russian steppe. At a village market in Rostov Oblast, women wearing headscarves and *abayat* chat in the shade of stalls piled high with watermelons and cabbages. They are Caucasian immigrants, some of whom hail from the occupied Georgian territories. The village is a short drive from Elista, the capital of the Russian republic of Kalmykia, home to the predominantly Buddhist Kalmyks who came here from Siberia in the seventeenth century (and reprised the journey in the 1950s after Stalin banished them). Russia has also received a huge injection of Ukrainians since the beginning of the war, although Ukraine claims that most of them were taken there against their will. Ukraine has witnessed a mass exodus, of course, as well as internal displacement. On the steppes north of the Black Sea, Bronze Age kurgans – the only blips in the horizontal – are being used for defence.

People are moving around the Black Sea today for the same reasons they did six thousand years ago: trade, war, climate change. The region seems particularly febrile in the third decade of the third millennium CE, but it's hard to think of a time when it was entirely settled and peaceful. And it's far from unique in that way. People are on the move somewhere in the world all the time. The proportion of the global population defined as international migrants has remained stable since 1960, at about three per cent.[5] Refugees constitute a special category of these. Their numbers are more variable, reflecting short-term crises, but on average they account for a small minority – ten per cent – of that three per cent.

Everywhere you turn today, you hear predictions that a tidal wave of refugees is heading for the world's more prosperous parts as the climate crisis – induced, this time, by human activities –

renders other regions uninhabitable. The predictions sometimes come with warnings attached, that the cultures and languages of the receiving countries are under threat, and with calls to raise the border fences ever higher. The climate crisis is real and its effects are already being felt, but scholars are divided as to the impact these will have on migration. Dutch sociologist Hein de Haas says there will be no tidal wave. The dire predictions are based on a crude totting up of the Earth's surface area likely to be adversely affected by climate change, and an assumption that the inhabitants of those areas will all head for Europe, North America, Australia or the richer parts of Asia. But de Haas believes that the crisis is more likely to trap the most vulnerable where they are. Some people will travel a long way, and stay where they end up, but because long-distance migration requires considerable resources, most will stray much less far and return home when they can – as they did in Europe in the past, and as they are doing in the Black Sea region today. Others see large pulses of migration as a distinct possibility, not because of climate change itself, but because of its social and economic consequences – loss of land, shortage of water and, eventually, wars over these.

However the crisis unfolds, the world is unlikely to stay the same, linguistically speaking. Only ten years ago, prehistorians thought that the Indo-European languages had surfed a tsunami of migrants to inundate Europe and parts of Asia. Today, the consensus is that those languages came to dominate mainly as a result of the small, temporary movements of people over time – the kind of displacements that de Haas foresees more of as climate change intensifies. Even if we can't predict the linguistic impact of these in any given place, overall that impact could be profound. We stand today in a familiar linguistic landscape, fretting over the languages we're losing, but we are not talking about the potentially far greater changes coming down the line.

It's a difficult subject to discuss. Understandably, a lot of people find it unnerving to think of their languages being transformed by the arrival of new ones, or even dying. But underpinning that fear is a false impression that they have remained static in the past. Throughout humanity's long existence, languages have never ceased to absorb and change each other. The version of a language that is regarded as standard is often the most changed of all, with respect to a common ancestor. Why? Because it is usually the one spoken in the capital or by an elite, and history tends to be more eventful in places and groups that concentrate power. A very funny illustration of this is the scene in the cult French film *Bienvenue chez les Ch'tis* (*Welcome to the Sticks*), where two men's differing pronunciation of the word for 'dog' causes mutual bewilderment. The northerner's *kien* preserves the hard *c* of Latin *canis*, while the standard French-speaker's *chien* does not.

Europe has promoted monolingualism ever since it embraced the philosophy of the nation-state in the eighteenth century, and from that time on languages and dialects other than the national ones were repressed. Before that, however, Europe was as multilingual as the rest of the world, and as our species for most of its past. ('I have three hearts,' wrote the Roman poet Ennius, who was fluent in Latin, Oscan and Greek.) Often it was those institutions now regarded as models of linguistic propriety that drove change, not least the monarchy. Richard the Lionheart, the great English king who ruled in the twelfth century, probably could not speak English. His mother tongue was Occitan. Elizabeth the First was particularly sensitive to new linguistic usages around her, and regularly ploughed them back into her own speech. She helped English do away with the double negative ('I did not do nothing') and replace 'ye' with 'you'.

Some inkling of what's to come might be the 'multiethnolects' that linguists have been observing excitedly for decades now: versions of the national language spoken by young people, mainly the children of immigrants, in multilingual, working-class urban

CONCLUSION

neighbourhoods. Kiezdeutsch in Germany, Straattaal in the Netherlands and Multicultural London English are examples. Multiethnolects exist alongside other spoken forms of the language, including those regional varieties that have survived, and act as conduits for linguistic innovation. Because they are often stigmatised they tend to be short-lived, but a change of circumstances could cause any one of them to stabilise and expand, even – one day – to become the standard form.

The new tools of archaeology and genetics have opened our eyes to our past. Migration has been a constant, 'indigenous' is relative. Ten thousand years of human displacement have shrunk the genetic distance between populations to the point where ethnic divisions are losing meaning. The desire to belong is as strong as ever, and as it becomes harder to see the difference between 'them' and 'us', linguistic and cultural boundaries are being guarded more jealously. Language is becoming a battleground in the identity wars, and preserving our linguistic 'purity' a justification used by those who want to raise walls. Unfortunately for them, the most successful language the world ever knew was a hybrid trafficked by migrants. It changed as it went, and when it stopped changing, it died.

The past is a lighthouse, not a port.†

---

† Russian proverb

Figures

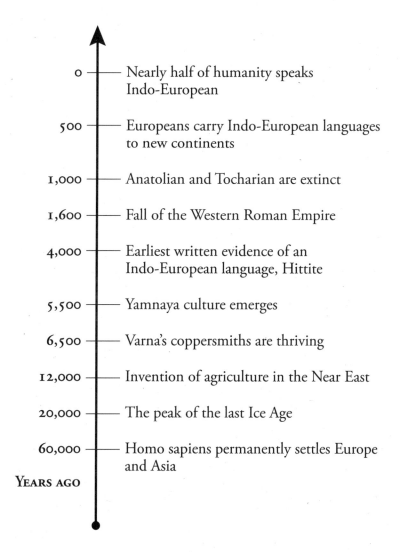

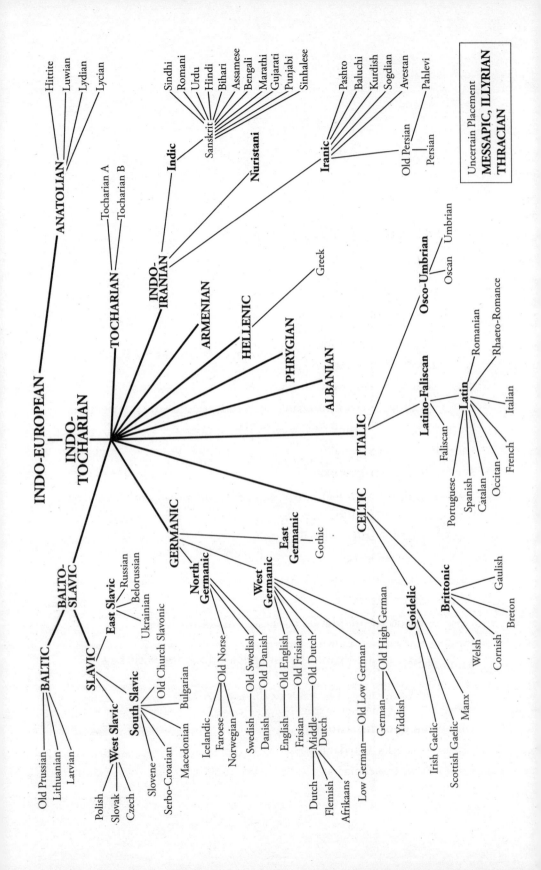

ACKNOWLEDGEMENTS

It was a conversation under a pagoda on the campus of the Institute of American Indian Arts in Santa Fe, New Mexico, in June 2022, that marked the birth of this project, and for that I have to thank language scientist Annika Tjuka. Annika was enrolled in the Santa Fe Institute's Complex Systems Summer School, as was I, and though we weren't talking about the Indo-European languages specifically, the passion she showed for her subject – in that thoroughly multidisciplinary environment – caused something to crystallise in my mind. So my first thanks go to Annika and to the Santa Fe Institute, which expanded all our minds that summer.

Many people gave generously of their time and knowledge in the making of this book. Most are named in the text, but not all, so I'd like to express my gratitude to others who conveyed their insight, erudition or practical advice when it was needed: Bruno Batllou and Pascale Landi of the Musée de la Romanité in Nîmes; Thomas Bak of the University of Edinburgh; Bruno Barbier, who shared a curious tale about a 2CV; tour guide Branimir Belchev from Varna; Zina Garces, a teacher of Russian literature; Annie Monneraie Goarin, a former cultural attaché in Armenia; archaeologists Bisserka Gaydarska and John Chapman of the University of Durham; molecular anthropologist Wolfgang Haak of the Max

Planck Institute for Evolutionary Anthropology; sociolinguist Raymond Hickey of the University of Duisburg and Essen; former language teacher Laura M. Keith; retired archaeologist Raymond Lanfranchi and his wife Catherine, a retired teacher; Radosław Liwoch of the Kraków Archaeological Museum; linguist Simon Poulsen of the University of Copenhagen; Corinna Salomon of the University of Vienna, an expert on runes; archaeologist Mykhailo Videiko of Borys Grinchenko Kyiv University; linguist Thomas Wier of the Free University of Tbilisi; and Mohammad Yahaghi, who teaches Persian literature in Mashhad, Iran. If the book contains any errors, responsibility for them is, of course, exclusively mine.

The Calouste Gulbenkian Foundation provided me with a grant that made it possible for me to visit Armenia, where I was fortunate enough to have Arpine Avetisyan as my guide. Ruben Badalyan showed me the Bronze Age collection at the State History Museum in Yerevan; Stéphane Deschamps took time out from his excavations at Erebuni to share his thinking on the Iron Age kingdom of Urartu; and Nazénie Garibian and Israël Lévi of the Matenadaran in Yerevan cut through some red tape to provide me with a document I needed.

Bogdan Draganski and Zornitsa Karamihova helped me organise my travels in Bulgaria, and Vladimir Slavchev answered my many questions while I was there, never baulking even when I dragged him down a rabbit hole the shape of a cow's thigh bone. Volker Heyd, Marianna Bálint and János Dani took a leap of faith when they let me observe their excavations in the Hortobágy National Park in June 2023. And when I was crippled by a migraine in the middle of the puszta, they could not have been kinder. My apologies go to them and to their team's chief excavator Bianca Preda-Bălănică for the liability that I temporarily became. At Boğazkale in Türkiye I had an excellent guide in Atila Ertugrul, who had assisted in the excavation of Hattusha since he was a child, and seen a few eminent archaeologists come and go. It was

ACKNOWLEDGEMENTS

all I could do to keep up with him, as he leapt through those vertiginous ruins like a mountain goat.

I wasn't sure, at the beginning of this project, if I would go to Russia. Plenty of organisations were boycotting it at the time, and I attended several conferences from which Russian academics had been excluded. I understand the reasons for this, and even agree with them. I thought about it deeply before deciding that to write a book about Proto-Indo-European, I had to visit the lands where it was spoken and meet one of the leading experts on its speakers. The appalling fact of the war only made that more obvious: what happened there in the Bronze Age is relevant to what is happening there now.

Needless to say it was not easy for a foreigner to travel in Russia in the summer of 2023, and without the indomitable Natalia Shishlina it would not have been possible. I would like to thank her and her sister Lilia, along with Vera Kuznetsova and her husband Andrey, Idris Idrisov, the rest of Natalia's team in Remontnoye and the mayor of that village, Anatoly Petrovitch Pustovetov, for making my trip so productive and, frankly, unforgettable. As Vera knows, I discovered my own shibboleth in the southern Russian steppe. Ekaterina Boldyreva and Maria Alexandrenkova kindly showed me around the State History Museum in Moscow. Linguist Alexei Kassian of the Russian Presidential Academy of National Economy and Public Administration in Moscow went out of his way to help, though practicalities got in the way of our meeting.

Guus Kroonen of Leiden University, one of the historical linguists and Indo-European experts who was quickest to embrace the ancient DNA revolution, patiently explained things to me on a number of occasions. Jay Jasanoff of Harvard University commented on sections of the book and dissected an extract from George Orwell's *1984*, even though I spelled his name wrong (I haven't now). Birgit Olsen of the University of Copenhagen commented helpfully on another section, and included me in a memorable dinner she organised in Yerevan. David Anthony and Jim Mallory showed great intellectual generosity, given that they haven't finished writing on this topic themselves. Alexey Nikitin answered my many questions in a faultlessly even-handed manner at a time of great personal distress, while the Russian regime waged war on the land of his birth.

Thanks go finally to my friends Sophie and Franck Fatalot, who offered me a place to write when my home was a building site, and to the village of Saint-Hippolyte-de-Caton, which offered me another. Thanks to my friends Kate and Ed Douglas for their advice and encouragement, especially to Kate who read a first draft and, with her expert storyteller's eye, highlighted some retrospectively obvious room for improvement. Thanks to my agent, Will Francis, for being enthused by the idea from the start, though we hadn't previously worked together, and to my excellent editors – especially Eva Hodgkin at William Collins – who helped turn it into a book, improving it again in the process. Last but by no means least, bottomless thanks to the majordomo, as the poets style him: my very own lionheart, Richard Frackowiak.

BIBLIOGRAPHY

This is not a textbook and I don't want to overload readers with references, but it's obviously important that I back up my arguments and give credit where it's due, so here are the essential references for each part (cited once only for each, though they may be relevant to more than one chapter) prefaced by my recommended reading list for anyone who wishes to go deeper, listed alphabetically in both cases. The general reading list is far from exhaustive, and some of the books I've included are more technical than others. Readers should also be aware that in a fast-moving field some of their contents are likely to be out of date.

GENERAL READING

Anthony D. W. 2007. *The Horse, the Wheel, and Language: How Bronze-Age Riders from the Eurasian Steppes Shaped the Modern World*. Princeton, New Jersey: Princeton University Press.

Demoule J.-P. 2023. *The Indo-Europeans: Archaeology, Language, Race, and the Search for the Origins of the West*. Translated by Rhoda Cronin-Allanic. New York: Oxford University Press.

Evans N. 2022. *Words of Wonder: Endangered Languages and What They Tell Us*. Oxford: Wiley-Blackwell. 2nd edn.

Fortson B. W. IV. 2010. *Indo-European Language and Culture: An Introduction*. Oxford: Blackwell. 2nd edn.

Krause J. and Trappe T. 2021. *A Short History of Humanity: How Migration Made Us Who We Are*. Translated by Caroline Waight. London: W. H. Allen.

Kristiansen K., Kroonen G. and Willerslev E. (eds). 2023. *The Indo-European Puzzle Revisited: Integrating Archaeology, Genetics, and Linguistics*. Cambridge: Cambridge University Press.

Mallory J. P. and Adams D. Q. 2006. *The Oxford Introduction to Proto-Indo-European and the Proto-Indo-European World*. New York: Oxford University Press.

Olander T. 2022. *The Indo-European Language Family: A Phylogenetic Perspective*. Cambridge: Cambridge University Press.

Ostler N. 2005. *Empires of the Word: A Language History of the World*. London: HarperCollins.

Reich D. 2018. *Who We Are and How We Got Here: Ancient DNA and the New Science of the Human Past*. New York: Pantheon.

References

Introduction: Ariomania

Barbieri C. et al. 2022. A global analysis of matches and mismatches between human genetic and linguistic histories. *Proceedings of the National Academy of Sciences* 119:47; e2122084119. https://doi.org/10.1073/pnas.2122084119

BBC News. Welsh woman on bus shuts down racist who told Muslim passenger to 'speak English'. 21 June 2016. https://www.bbc.com/news/newsbeat-36580448

Berge A. 2017. Subsistence terms in Unangam Tunuu (Aleut). In: *Language Dispersal Beyond Farming*. Robbeets M. and Savelyev A. (eds). Amsterdam: John Benjamins. 3:47–73. https://doi.org/10.1075/z.215.03ber

BIBLIOGRAPHY

Callaway E. 'Truly gobsmacked': Ancient-human genome count surpasses 10,000. *Nature* 617:20. 24 April 2023. https://www.nature.com/articles/d41586-023-01403-4#:~:text=In%20the%2013%20years%20since,Medical%20School%20in%20Boston%2C%20Massachusetts.

Dediu D. and Levinson S. C. 2013. On the antiquity of language: the reinterpretation of Neandertal linguistic capacities and its consequences. *Frontiers in Psychology* 4:397. https://doi.org/10.3389/fpsyg.2013.00397

Evans N. 2017. Did language evolve in multilingual settings? *Biology & Philosophy* 32:905–33. https://doi.org/10.1007/s10539-018-9609-3

Fabbro F., Fabbro A. and Crescenti C. 2022. The nature and function of languages. *Languages* 7:303. https://www.mdpi.com/2226-471X/7/4/303

Fisher M. A revealing map of the most and least ethnically diverse countries. *The Washington Post*. 16 May 2013. https://www.washingtonpost.com/news/worldviews/wp/2013/05/16/a-revealing-map-of-the-worlds-most-and-least-ethnically-diverse-countries/

Herodotus. *The Histories*. 2003. Translated by de Sélincourt A., Marincola J. M. and Radice B. (eds). London: Penguin. Rev. edn.

Hua X. et al. 2019. The ecological drivers of variation in global language diversity. *Nature Communications* 10:2047. https://doi.org/10.1038/s41467-019-09842-2

Leroux M. et al. 2023. Call combinations and compositional processing in wild chimpanzees. *Nature Communications* 14:2225. https://doi.org/10.1038/s41467-023-37816-y

Nichols J. 2015. Types of spread zones: open and closed, horizontal and vertical. In: *Language Structure and Environment: Social, Cultural, and Natural Factors*. De Busser R. and LaPolla R. J. (eds). Cognitive Linguistic Studies in Cultural Contexts 6. 10:261–86. https://doi.org/10.1075/clscc.6.10nic

Pagel M. 2000. The history, rate and pattern of world linguistic evolution. In: *The Evolutionary Emergence of Language: Social Function and the Origins of Linguistic Form*. Knight C., Studdert-Kennedy M. and Hurford J. (eds). Cambridge: Cambridge University Press. 391–416.

Planer R. J. and Sterelny K. 2021. *From Signal to Symbol: The Evolution of Language*. Cambridge, Massachusetts: The MIT Press.

Sherratt A. and Sherratt S. 1988. The archaeology of Indo-European: an alternative view. *Antiquity* 62(236):584–95. https://doi.org/10.1017/S0003598X00074755

1: Genesis

Baker P. 2002. *Polari: The Lost Language of Gay Men*. London: Routledge.

Braarvig J. and Geller M. J. 2018. *Multilingualism, Lingua Franca and Lingua Sacra*. Berlin: Max Planck Research Library for the History and Development of Knowledge, Studies 10. https://doi.org/10.34663/9783945561133-00

Burmeister S. 2017. Early wagons in Eurasia: disentangling an enigmatic innovation. In: *Appropriating Innovations. Entangled Knowledge in Eurasia, 5000–1500 BCE*. Stockhammer P. and Maran J. (eds). Oxford: Oxbow. 69–77.

Cameron C. M. 2016. *Captives: How Stolen People Changed the World*. Lincoln, Nebraska: University of Nebraska Press.

Chintalapati M., Patterson N. and Moorjani P. 2022. The spatiotemporal patterns of major human admixture events during the European Holocene. *eLife* 11:e77625. https://doi.org/10.7554/eLife.77625

D'Iakonov I. M. 1984. On the Original Home of the Speakers of Indo-European. *Soviet Anthropology and Archeology* 23(2):5–77. https://doi.org/10.2753/AAE1061-195923025

Gamkrelidze T. V. and Ivanov V. V. 1990. The Early History of Indo-European Languages. *Scientific American* 262(3):110–17. https://www.jstor.org/stable/24996796

Gaydarska B., Beavan N. and Slavchev V. 2022. Lifeway interpretations from ancient diet in the Varna cemetery. *Oxford Journal of Archaeology* 41(1):22–41. https://doi.org/10.1111/ojoa.12236

Gronenborn D. et al. 2020. A later fifth-millennium cal BC tumulus at Hofheim-Kapellenberg, Germany. *Antiquity* 94(375):e14. https://doi.org/10.15184/aqy.2020.79

Immel A. et al. 2020. Gene-flow from steppe individuals into Cucuteni-Trypillia associated populations indicates long-standing contacts and gradual admixture. *Scientific Reports* 10:4253. https://doi.org/10.1038/s41598-020-61190-0

Ivanova M. 2012. Perilous waters: early maritime trade along the western coast of the Black Sea (fifth millennium BC). *Oxford Journal of Archaeology* 31(4):339–65. https://doi.org/10.1111/j.1468-0092.2012.00392.x

Kondor D. et al. 2023. Explaining population booms and busts in Mid-Holocene Europe. *Scientific Reports* 13:9301. https://doi.org/10.1038/s41598-023-35920-z

Küchelmann H. C. and Zidarov P. 2005. Let's skate together! Skating on bones in the past and today. In: *From Hooves to Horns, From Mollusc to Mammoth: Manufacture and Use of Bone Artefacts from Prehistoric Times to the Present*. Luik H., Choyke A. M., Batey C. E. and Lõugas L. (eds). Proceedings of the 4th meeting of the ICAZ Worked Bone Research Group at Tallinn, Estonia, 26–31 August 2003.

Lazaridis I., Alpaslan-Roodenberg S. et al. 2022. The genetic history of the Southern Arc: A bridge between West Asia and Europe. *Science* 377:eabm4247. https://www.science.org/doi/10.1126/science.abm4247

Mallory J. P. 2014. Indo-European dispersals and the Eurasian steppe. In: *Reconfiguring the Silk Road: New Research on East-West Exchange in Antiquity*. Mair V. H. and Hickman J. (eds). Philadelphia, Pennsylvania: University of Pennsylvania Press. 73–88. https://doi.org/10.9783/9781934536698.73

Mathieson I. et al. 2018. The genomic history of southeastern Europe. *Nature* 555:197–203. https://doi.org/10.1038/nature25778

Mattila T. M. et al. 2023. Genetic continuity, isolation, and gene flow in Stone Age Central and Eastern Europe. *Communications Biology*. 6(1):793. https://doi.org/10.1038/s42003-023-05131-3

Nikitin A. G., Ivanova S. 2023. Long-distance exchanges along the Black Sea coast in the Eneolithic and the steppe genetic ancestry problem. In: *Steppe Transmissions*. Archaeolingua (Budapest), volume 45: The Yamnaya Impact on Prehistoric Europe. Preda-Bălănică B. and Ahola M. (eds). 9–27. https://doi.org/10.33774/coe-2022-7m315

Nikitin A. G. et al. 2023. Interactions between Trypillian farmers and North Pontic forager-pastoralists in Eneolithic central Ukraine. *PLoS ONE* 18(6):e0285449. https://doi.org/10.1371/journal.pone.0285449

Nikolov V. 2018. Social dimensions of salt in the later prehistory of the eastern Balkans. In: *Social Dimensions of Food in the Prehistoric Balkans*. Ivanova M. et al. (eds). Oxford: Oxbow Books. 215–20. https://doi.org/10.2307/j.ctvh1dsx3

Okrostsvaridze A., Gagnidze N. and Akimidze K. 2016. A modern field investigation of the mythical 'gold sands' of the ancient Colchis Kingdom and 'Golden Fleece' phenomena. *Quaternary International* 409(A):61–9. https://doi.org/10.1016/j.quaint.2014.07.064

Spinney L. How farmers conquered Europe. *Scientific American* 60–67. July 2020. https://www.scientificamerican.com/article/when-the-first-farmers-arrived-in-europe-inequality-evolved/

Spinney L. The rise and fall of the mysterious culture that invented civilisation. *New Scientist* 44–7. 27 February 2021. https://www.newscientist.com/article/mg24933230-900-the-rise-and-fall-of-the-mysterious-culture-that-invented-civilisation/

Trifonov V. et al. 2019. A 5000-year-old souslik fur garment from an elite megalithic tomb in the North Caucasus, Maykop culture. *Paléorient* 45(1):69–80. https://www.jstor.org/stable/10.2307/26742475

Trifonov V., Petrov D. and Savelieva L. 2022. Party like a Sumerian: reinterpreting the 'sceptres' from the Maikop kurgan. *Antiquity* 96(385):67–84. https://doi.org/10.15184/aqy.2021.22

Wang C.-C. et al. 2019. Ancient human genome-wide data from a 3000-year interval in the Caucasus corresponds with eco-geographic regions. *Nature Communications* 10:590. https://doi.org/10.1038/s41467-018-08220-8

Wilkin S. et al. 2023. Curated cauldrons: Preserved proteins from early copper-alloy vessels illuminate feasting practices in the Caucasian steppe. *iScience.* https://doi.org/10.1016/j.isci.2023.107482

Yanko-Hombach V., Gilbert A. S., Panin N., Dolukhanov P. M (eds). 2007. *The Black Sea Flood Question: Changes in Coastline, Climate and Human Settlement.* Dordrecht: Springer. https://doi.org/10.1007/978-1-4020-5302-3

2: Sacred Spring

Anthony D. W. and Ringe D. 2015. The Indo-European Homeland from Linguistic and Archaeological Perspectives. *Annual Review of Linguistics.* 1:199–219. https://doi.org/10.1146/annurev-linguist-030514-124812

Anthony D. W. et al. 2022. The Eneolithic cemetery at Khvalynsk on the Volga River. *Praehistorische Zeitschrift* 97(1):22–67. https://doi.org/10.1515/pz-2022-2034

Anthony D. W. 2023. Ten constraints that limit the late PIE homeland to the steppes. In: *Proceedings of the 33rd Annual UCLA Indo-European Conference.* Goldstein D. M., Jamison S. W. and Yates A. D. (eds). Hamburg: Buske. 1–25.

Clackson J. 2007. *Indo-European Linguistics: An Introduction.* Cambridge: Cambridge University Press.

D'Huy J. 2020. *Cosmogonies: La préhistoire des mythes.* Paris: La Découverte.

Höfler S., Ginevra R. and Olsen B. 2024. *Power, Gender, and Mobility: Aspects of Indo-European Society.* Copenhagen: Museum Tusculanum.

Jones J. Becoming a centaur. *Aeon.* 14 January 2022. https://aeon.co/essays/horse-human-cooperation-is-a-neurobiological-miracle

Kroonen G. et al. 2022. Indo-European cereal terminology suggests a Northwest Pontic homeland for the core Indo-European languages. *PLoS ONE* 17(10):e0275744. https://doi.org/10.1371/journal.pone.0275744

Lazaridis I., Alpaslan-Roodenberg S. et al. 2022. A genetic probe into the ancient and medieval history of Southern Europe and West Asia. *Science* 377:940–51. https://doi.org/10.1126/science.abq0755

Lazaridis I., Patterson N., Anthony D. et al. 2024. The genetic origin of the Indo-Europeans. *bioRxiv* (preprint). https://doi.org/10.1101/2024.04.17.589597

Librado P. et al. 2021. The origins and spread of domestic horses from the Western Eurasian steppes. *Nature* 598:634–40. https://doi.org/10.1038/s41586-021-04018-9

Magazine, *The Horse*. 2018. Federico Grisone – was he really such a baddie? https://www.horsemagazine.com/thm/2015/03/federico-grisone-was-he-really-such-a-baddie/

Mathieson I. et al. 2015. Genome-wide patterns of selection in 230 ancient Eurasians. *Nature* 528:499–503. https://doi.org/10.1038/nature16152

McPartland J. M. and Hegman W. 2018. Cannabis utilization and diffusion patterns in prehistoric Europe. *Vegetation History and Archaeobotany* 27(4):627–34. https://doi.org/10.1007/s00334-017-0646-7

Nikitin A. G., Lazaridis I. et al. 2024. A genomic history of the North Pontic Region from the Neolithic to the Bronze Age. *bioRxiv* (preprint). https://doi.org/10.1101/2024.04.17.589600

Scott A. et al. 2022. Emergence and intensification of dairying in the Caucasus and Eurasian steppes. *Nature Ecology & Evolution* 6:813–22. https://doi.org/10.1038/s41559-022-01701-6

Shishlina, N. 2008. *Reconstruction of the Bronze Age of the Caspian Steppes: Life Styles and Life Ways of Pastoral Nomads*. Oxford: Archaeopress.

Trautmann M. et al. 2023. First bioanthropological evidence for Yamnaya horsemanship. *ScienceAdvances* 9:eade2451. https://doi.org/10.1126/sciadv.ade2451

Watkins C. 1995. *How to Kill a Dragon: Aspects of Indo-European Poetics*. Oxford: Oxford University Press.

Wilkin S. et al. 2021. Dairying enabled Early Bronze Age Yamnaya steppe expansions. *Nature* 598:629–33. https://doi.org/10.1038/s41586-021-03798-4

3: First Among Equals

Allentoft M. E. et al. 2015. Population genomics of Bronze Age Eurasia. *Nature* 522:167–72. https://doi.org/10.1038/nature14507

Bryce T. 2005. *The Kingdom of the Hittites*. Oxford: Oxford University Press. https://doi.org/10.1093/acprof:oso/9780199281329.001.0001

Buck C. D. 1920. Hittite an Indo-European language? *Classical Philology* XV:184–92. https://www.jstor.org/stable/263436

Collins B. J. 1998. Ḫattušili I, The Lion King. *Journal of Cuneiform Studies* 50:15–20. https://doi.org/10.2307/1360028

De Barros Damgaard P. et al. 2018. The first horse herders and the impact of early Bronze Age steppe expansions into Asia. *Science* 360:eaar7711. https://www.science.org/doi/10.1126/science.aar7711

Elster E. S. 2015. In Memoriam: Marija Gimbutas: old Europe, goddesses and gods, and the transformation of culture. *Backdirt: Annual Review of The Cotsen Institute of Archaeology at UCLA*. 94–102.

Gimbutas M. 1963. The Indo-Europeans: archeological problems. *American Anthropologist* 65(4):815–36. https://www.jstor.org/stable/668932

Ginevra R. 2019. Myths of non-functioning fertility deities in Hittite and core Indo-European. In: *Dispersals and Diversification: Linguistic and Archaeological Perspectives on the Early Stages of Indo-European*. Serangeli M. and Olander T. (eds). Leiden: Brill. 4:106–29. https://doi.org/10.1163/9789004416192_006

Goedegebuure P. M. 2008. Central Anatolian languages and language communities in the colony period: a Luwian-Hattian symbiosis and the independent Hittites. In: *Anatolia and the Jazira during the Old Assyrian Period*. Dercksen J. G. (ed). Leuven: Peeters. 137–80.

Graeber D. and Wengrow D. 2021. *The Dawn of Everything: A New History of Humanity*. London: Penguin.

Haak W. et al. 2015. Massive migration from the steppe was a source for Indo-European languages in Europe. *Nature* 522(7555):207–11. https://doi.org/10.1038/nature14317

Heggarty P. et al. 2023. Language trees with sampled ancestors support a hybrid model for the origin of Indo-European languages. *Science* 381:6656. https://doi.org/10.1126/science.abg0818

Jones E. R. et al. 2015. Upper Palaeolithic genomes reveal deep roots of modern Eurasians. *Nature Communications*. 16(6):8912. https://doi.org/10.1038/ncomms9912

Julius-Maximilians-Universität Würzburg. New Indo-European language discovered. 21 September 2023. https://www.uni-wuerzburg.de/en/news-and-events/news/detail/news/new-indo-european-language-discovered/

Kristiansen K. 2020. The archaeology of Proto-Indo-European and Proto-Anatolian: locating the split. In: *Dispersals and Diversification Linguistic and Archaeological Perspectives on the Early Stages of Indo-European*. Serangeli M. and Olander T. (eds). Leiden: Brill. 7:157–65. https://doi.org/10.1163/9789004416192_009

Kroonen G., Barjamovic G. and Peyrot M. Linguistic supplement to Damgaard et al. (2018): Early Indo-European languages, Anatolian, Tocharian and Indo-Iranian. https://doi.org/10.5281/zenodo.1240524

Mallory J. P. 1989. *In Search of the Indo-Europeans: Language, Archaeology, and Myth*. London: Thames & Hudson.

Melchert H. C. 1991. Death and the Hittite king. In: *Perspectives on Indo-European Language, Culture and Religion: Studies in Honor of Edgar C. Polomé*. Pearson R. (ed). 1:182–8.

Pereltsvaig A. and Lewis M. W. 2015. *The Indo-European Controversy: Facts and Fallacies in Historical Linguistics*. Cambridge: Cambridge University Press.

Renfrew C. 1987. *Archaeology and Language: The Puzzle of Indo-European Origins*. London: Cape.

Sturtevant E. H. 1926. On the position of Hittite among the Indo-European languages. *Language* 2:25–34. https://doi.org/10.2307/408784

Watkins C. 2004. The third donkey: origin legends and some hidden Indo-European themes. In: *Indo-European Perspectives. Studies in Honour of Anna Morpurgo Davies*. Oxford: Oxford University Press. 65–79.

Yakubovich I. 2010. Sociolinguistics of the Luvian language. In: *Studies in Indo-European Languages and Linguistics*, volume 2. Leiden: Brill.

4: Over The Range

Allentoft M. E., Sikora M., Refoyo-Martínez A. et al. 2024. Population genomics of post-glacial western Eurasia. *Nature* 625:301–11. https://doi.org/10.1038/s41586-023-06865-0

Aurel Stein M. 1909. Explorations in Central Asia, 1906–8. *The Geographical Journal* 34(1):5–36. https://www.jstor.org/stable/1777985

Betts A., Jia P. and Abuduresule I. 2019. A new hypothesis for early Bronze Age cultural diversity in Xinjiang, China. *Archaeological Research in Asia* 17:204–13. https://doi.org/10.1016/j.ara.2018.04.001

Chen K.-T. and Hiebert F. T. 1995. The late prehistory of Xinjiang in relation to its neighbors. *Journal of World Prehistory* 9(2):243–300. https://www.jstor.org/stable/25801077

Cope T. 13 January 2014. On the trail of Genghis Khan. *Nat Geo Live*. https://www.youtube.com/watch?v=FBFGLm5HwXY

Frachetti M. D. 2012. Multiregional emergence of mobile pastoralism and nonuniform institutional complexity across Eurasia. *Current Anthropology* 53(1):2–38. https://www.jstor.org/stable/10.1086/663692

Gasparini M. 2014. A mathemetic expression of art: Sino-Iranian and Uighur textile interaction and the Turfan Textile Collection in Berlin. *The Journal of Transcultural Studies* 5(1):134–63. https://doi.org/10.11588/ts.2014.1.12313

Kuz'mina E. E. 2007. The origin of the Indo-Iranians. In: *Leiden Indo-European Etymological Dictionary Series*, volume 3. Leiden: Brill. https://doi.org/10.1163/ej.9789004160545.i-763

Mallory J. P. 2015. The problem of Tocharian origins: an archaeological perspective. In: *Sino-Platonic Papers*. Mair V. H. (ed). 259.

Mallory J. P. and Mair V. H. 2000. *The Tarim Mummies: Ancient China and the Mystery of the Earliest Peoples From the West*. London: Thames & Hudson.

Peyrot M. 2018. Tocharian B *etswe* 'mule' and Eastern East Iranian. In: *Farnah: Indo-Iranian and Indo-European Studies in Honor of Sasha Lubotsky*. Ann Arbor, Michigan: Beech Stave Press.

Peyrot M. 2019. The deviant typological profile of the Tocharian branch of Indo-European may be due to Uralic substrate influence. *Indo-European Linguistics* 7(1):72–121. https://doi.org/10.1163/22125892-00701007

Pillai K. D. 2023. The hybrid origin of Brāhmī script from Aramaic, Phoenician and Greek letters. *Indi@logs* 10:93–122. https://doi.org/10.5565/rev/indialogs.213

Ringbauer H., Huang Y., Akbari A. et al. 2024. Accurate detection of identity-by-descent segments in human ancient DNA. *Nature Genetics* 56:143–51. https://doi.org/10.1038/s41588-023-01582-w

Struck J. et al. 2022. Central Mongolian lake sediments reveal new insights on climate change and equestrian empires in the Eastern Steppes. *Scientific Reports* 12:2829. https://doi.org/10.1038/s41598-022-06659-w

Wang C.-C. et al. 2021. Genomic insights into the formation of human populations in East Asia. *Nature* 591:413–19. https://doi.org/10.1038/s41586-021-03336-2

Zhang F. et al. 2021. The genomic origins of the Bronze Age Tarim Basin mummies. *Nature* 599:256–61. https://doi.org/10.1038/s41586-021-04052-7

5: Lark Rising

Armit I. and Reich D. 2021. The return of the Beaker folk? Rethinking migration and population change in British prehistory. *Antiquity* 95(384):1464–77 https://doi.org/10.15184/aqy.2021.129

Barrie W. et al. 2024. Elevated genetic risk for multiple sclerosis emerged in steppe pastoralist populations. *Nature* 625:321–8. https://doi.org/10.1038/s41586-023-06618-z

Blasi D. E. et al. 2019. Human sound systems are shaped by post-Neolithic changes in bite configuration. *Science* 363(6432):eaav3218. https://doi.org/10.1126/science.aav3218

Booth T. J. et al. 2021. Tales from the supplementary information: ancestry change in Chalcolithic–Early Bronze Age Britain was gradual with varied kinship organization. *Cambridge Archaeological Journal* 31(3):379–400. https://doi.org/10.1017/S0959774321000019

Cassidy L. M. et al. 2016. Neolithic and Bronze Age migration to Ireland and establishment of the insular Atlantic genome. *Proceedings of the National Academy of Sciences* 113(2):368–73. https://doi.org/10.1073/pnas.1518445113

Goldberg A. et al. 2017. Ancient X chromosomes reveal contrasting sex bias in Neolithic and Bronze Age Eurasian migrations. *Proceedings of the National Academy of Sciences* 114(10):2657–62. https://doi.org/10.1073/pnas.1616392114

Gretzinger J. et al. 2022. The Anglo-Saxon migration and the formation of the early English gene pool. *Nature* 610:112–19. https://doi.org/10.1038/s41586-022-05247-2

Guyon L., Guez J., Toupance B. et al. 2024. Patrilineal segmentary systems provide a peaceful explanation for the post-Neolithic Y-chromosome bottleneck. *Nature Communications* 15:3243. https://doi.org/10.1038/s41467-024-47618-5

Heyd V. 2017. Kossinna's smile. *Antiquity* 91(356):348–59. https://doi.org/10.15184/aqy.2017.21

Hickey R. 2007. *Irish English: History and Present-day Forms.* Cambridge: Cambridge University Press.

Iversen R. and Kroonen G. 2017. Talking Neolithic: linguistic and archaeological perspectives on how Indo-European was implemented in southern Scandinavia. *American Journal of Archaeology* 121(4):511–25. https://doi.org/10.3764/aja.121.4.0511

Kassian A. et al. 2021. Rapid radiation of the inner Indo-European languages: an advanced approach to Indo-European lexicostatistics. *Linguistics* 59:949–79. https://doi.org/10.1515/ling-2020-0060

Knipper C. et al. 2017. Female exogamy and gene pool diversification at the transition from the Final Neolithic to the Early Bronze Age in central Europe. *Proceedings of the National Academy of Sciences* 114(38):10083–8. https://doi.org/10.1073/pnas.1706355114

Kristiansen K. 2018. The rise of Bronze Age peripheries and the expansion of international trade 1950–1100 BC. In: *Trade and Civilisation: Economic Networks and Cultural Ties, From Prehistory to the Early Modern Era*. Kristiansen K., Lindkvist T. and Myrdal J. (eds). Cambridge: Cambridge University Press. 87–112.

Kristiansen K. et al. 2017. Re-theorising mobility and the formation of culture and language among the Corded Ware Culture in Europe. *Antiquity* 91(356):334–47. https://doi.org/10.15184/aqy.2017.17

Kristiansen K. and Larsson T. 2005. *The Rise of Bronze Age Society: Travels, Transmissions, and Transformations*. Cambridge: Cambridge University Press.

Maier R. et al. 2023. On the limits of fitting complex models of population history to f-statistics. *eLife* 12:e85492. https://doi.org/10.7554/eLife.85492

Mallory J. P. 2016. *In Search of the Irish Dreamtime: Archaeology and Early Irish Literature*. London: Thames & Hudson.

Markale J. 1993. *The Celts: Uncovering the Mythic and Historic Origins of Western Culture*. Rochester, Vermont: Inner Traditions.

McColl H. et al. 2024. Steppe ancestry in western Eurasia and the spread of the Germanic languages. *bioRxiv* (preprint). https://doi.org/10.1101/2024.03.13.584607

McLaughlin T. R. 2020. An archaeology of Ireland for the Information Age. *Emania* 25:7–29.

Mittnik A. et al. 2019. Kinship-based social inequality in Bronze Age Europe. *Science* 366:731–4. https://doi.org/10.1126/science.aax6219

Olalde I. et al. 2018. The Beaker phenomenon and the genomic transformation of northwest Europe. *Nature* 555:190–96. https://doi.org/10.1038/nature25738

Orwell G. 2021. *Nineteen Eighty-Four*. London: Penguin Classics.

Papac L. et al. 2021. Dynamic changes in genomic and social structures in third millennium BCE central Europe. *Science Advances* 7:eabi6941. https://doi.org/10.1126/sciadv.abi6941

Parasayan O. et al. 2024. Late Neolithic collective burial reveals admixture dynamics during the third millennium BCE and the shaping of the European genome. *Science Advances* 10,eadl2468. https://doi.org/10.1126/sciadv.adl2468

Patterson N. et al. 2022. Large-scale migration into Britain during the Middle to Late Bronze Age. *Nature* 601:588–94. https://doi.org/10.1038/s41586-021-04287-4

Pinkstone J. The most violent group of people who ever lived. *Mail Online*. 29 March 2019. https://www.dailymail.co.uk/sciencetech/article-6865741/The-violent-group-people-lived.html

Posth C. et al. 2021. The origin and legacy of the Etruscans through a 2000-year archeogenomic time transect. *Science Advances* 7:eabi7673. https://doi.org/10.1126/sciadv.abi7673

Rio J., Quilodrán C. S. and Currat M. 2021. Spatially explicit paleogenomic simulations support cohabitation with limited admixture between Bronze Age Central European populations. *Communications Biology* 4:1163. https://doi.org/10.1038/s42003-021-02670-5

Saupe T. et al. 2021. Ancient genomes reveal structural shifts after the arrival of Steppe-related ancestry in the Italian Peninsula. *Current Biology* 31:2576–91. https://doi.org/10.1016/j.cub.2021.04.022

Scheidel W. 2008. Monogamy and polygyny in Greece, Rome, and World History. http://dx.doi.org/10.2139/ssrn.1214729

Schrijver P. 2014. *Language Contact and the Origins of the Germanic Languages*. New York and London: Routledge.

Sherratt A. 1987. Cups that cheered: the introduction of alcohol to prehistoric Europe. In: *Economy and Society in Prehistoric Europe* (1997). https://doi.org/10.1515/9781474472562-018

Sikora M. et al. 2023. The landscape of ancient human pathogens in Eurasia from the Stone Age to historical times. *bioRxiv* (preprint). https://doi.org/10.1101/2023.10.06.561165

Sims-Williams P. 2020. An alternative to 'Celtic from the East' and 'Celtic from the West'. *Cambridge Archaeological Journal* 30(3):511–29. https://doi.org/10.1017/S0959774320000098

Sjögren K.-G. et al. 2020. Kinship and social organization in Copper Age Europe. A cross-disciplinary analysis of archaeology, DNA, isotopes, and anthropology from two Bell Beaker cemeteries. *PLoS ONE* 15(11):e0241278. https://doi.org/10.1371/journal.pone.0241278

Stifter D. 2020. The early Celtic epigraphic evidence and early literacy in Germanic languages. *NOWELE* 73(1):123–52. https://doi.org/10.1075/nowele.00037.sti

Svizzero S. 2015. The collapse of the Únětice culture: economic explanation based on the 'Dutch disease'. *Czech Journal of Social Sciences, Business and Economics* 4(3):6–18.

Van Sluis P. 2022. Beekeeping in Celtic and Indo-European. *Studia Celtica* 56(1):1–28. https://doi.org/10.16922/SC.56.1

Wigman A. 2023. Unde Vēnistī?: The prehistory of Italic through its loanword lexicon. PhD thesis defended at Leiden University (unpublished).

6: The Wandering Horse

Beckwith C. I. 2023. *The Scythian Empire: Central Eurasia and the Birth of the Classical Age from Persia to China*. Princeton, New

Jersey: Princeton University Press.

Bergquist A. and Taylor T. 1987. The origin of the Gundestrup cauldron. *Antiquity* 61:10–24. https://doi.org/10.1017/S0003598X00072446

Bryant E. 2001. *The Quest for the Origins of Vedic Culture: The Indo-Aryan Migration Debate.* New York: Oxford University Press.

Chernykh E. N. et al. 2002. *Kargaly,* volume 1: Geological and geographical characteristics: history of discoveries, exploitation and investigations: archaeological sites. Chernykh E. N. (ed). Moscow: Languages of Slavonic Culture.

Dupré M.-C. and Pinçon B. 1997. *Métallurgie et Politique en Afrique centrale: Deux mille ans de vestiges sur les plateaux batéké. Gabon, Congo, Zaïre.* Paris: Karthala.

Graça da Silva S. and Tehrani J. J. 2016. Comparative phylogenetic analyses uncover the ancient roots of Indo-European folktales. *Royal Society Open Science* 3:150645. http://dx.doi.org/10.1098/rsos.150645

Grünthal R. et al. 2022. Drastic demographic events triggered the Uralic spread. *Diachronica* 39(4):490–524. https://doi.org/10.1075/dia.20038.gru

Heggarty P. 2015. Prehistory through language and archaeology. In: *The Routledge Handbook of Historical Linguistics.* Bowern C. and Evans B. (eds). Chapter 28. London: Routledge.

Kümmel M. J. 2017. Agricultural terms in Indo-Iranian. In: *Language Dispersal Beyond Farming.* Robbeets M. and Savelyev A. (eds). Amsterdam: John Benjamins. 12:275–90. https://doi.org/10.1075/z.215.12kum

Kupriyanova E. 2023. Burial rite of early Indo-European Bronze Age communities in southern Trans-Urals (Russia): a mirror of religion and society. Global Journal of Human-Social Science: D (History, Archaeology & Anthropology) 23(2): Version 1.0.

Librado P., Tressières G., Chauvey L. et al. 2024. Widespread horse-based mobility arose around 2,200 BCE in Eurasia. *Nature.* 631:819–825. https://doi.org/10.1038/s41586-024-07597-5

Narasimhan V. M. et al. 2019. The formation of human populations in South and Central Asia. *Science* 365:eaat7487. https://doi.org/10.1126/science.aat7487

Nichols J. 2021. The origin and dispersal of Uralic: distributional typological view. *Annual Review of Linguistics* 7:351–69. https://doi.org/10.1146/annurev-linguistics-011619-030405

Shinde V. et al. 2019. An ancient Harappan genome lacks ancestry from steppe pastoralists or Iranian farmers. *Cell* 179:1–7. https://doi.org/10.1016/j.cell.2019.08.048

Vuković K. 2023. *Wolves of Rome: The Lupercalia from Roman and Comparative Perspectives*. Berlin/Boston, Massachusetts: de Gruyter.

Witzel M. 2001. Autochthonous Aryans? The evidence from Old Indian and Iranian texts. *Electronic Journal of Vedic Studies* 7(3):1–93. http://dx.doi.org/10.11588/ejvs.2001.3.830

7: Northern Idyll

Carpenter H. 1981. *The Letters of J. R. R. Tolkien*. London: George Allen & Unwin.

French K. M. et al. 2024. Biomolecular evidence reveals mares and long-distance imported horses sacrificed by the last pagans in temperate Europe. *Science Advances* 10:eado3529. https://www.science.org/doi/10.1126/sciadv.ado3529

Gimbutas M. 1963. *The Balts*. London: Thames & Hudson.

Heather P. 2010. *Empires and Barbarians: Migration, Development and the Birth of Europe*. Basingstoke and Oxford: Pan Books.

Jakob A. 2024. A history of East Baltic through language contact. In: *Leiden Studies in Indo-European*, volume 24. Leiden: Brill.

Lang V. 2018. *Läänemeresoome tulemised: Finnic be-comings*. Tartu, Estonia: University of Tartu Press.

Larsson J. H. 2017. Sources for the early Baltic pantheon. In: *Language and Prehistory of the Indo-European Peoples: A Cross-disciplinary Perspective*. Hyllested A., Whitehead B. N., Olander T. and Olsen B.A. (eds). Copenhagen: Museum Tusculanum Press.

Larsson J. H. 2021. *In The Beginning Was The Word: Languages and Human Origins*. Stockholm: Makadam Förlag.

Mittnik A. et al. 2018. The genetic prehistory of the Baltic Sea region. *Nature Communications* 9:442. https://doi.org/10.1038/s41467-018-02825-9

Pereltsvaig A. 2011. Birch Bark letters. Languages of the world: blog of Asya Pereltsvaig. https://www.languagesoftheworld.info/historical-linguistics/birch-bark-letters-part-1.html

Pitt M. A. 1998. Phonological processes and the perception of phonotactically illegal consonant clusters. *Perception & Psychophysics* 60(6):941–51. https://doi.org/10.3758/bf03211930

Pronk T. and Pronk-Tiethoff S. 2018. Balto-Slavic agricultural terminology. In: *Talking Neolithic: Proceedings of the Workshop on Indo-European Origins Held at the Max Planck Institute for Evolutionary Anthropology, Leipzig, December 2–3, 2013*. Kroonen G., Mallory J. P. and Comrie B. (eds). 278–314.

Rębała K. et al. 2007. Y-STR variation among Slavs: evidence for the Slavic homeland in the middle Dnieper basin. *Journal of Human Genetics* 52:406–14. https://doi.org/10.1007/s10038-007-0125-6

Saag L. et al. 2019. The arrival of Siberian ancestry connecting the Eastern Baltic to Uralic speakers further east. *Current Biology* 29:1701–11. https://doi.org/10.1016/j.cub.2019.04.026

Schenker A. M. 1995. *The Dawn of Slavic: An Introduction to Slavic Philology*. New Haven, Connecticut and London: Yale University Press.

8: They Came From Steep Wilusa

Bobokhyan A. and Martirosyan-Olshansky K. 2022. Transformation of a sacred landscape around Lake Gilli, Armenia. *Backdirt: Annual Review of The Cotsen Institute of Archaeology at UCLA*. 16–27.

Durkin P. 2014. *Borrowed Words: A History of Loanwords in English*. Oxford: Oxford University Press.

Hovhannisyan A., Maisano Delser P. et al. 2024. Demographic history and genetic variation of the Armenian population. *The American Journal of Human Genetics* 112: 1–17. https://doi.org/10.1016/j.ajhg.2024.10.022

Jendraschek, G. 2019. Romance in contact with Basque. In: *Oxford Research Encyclopedia of Linguistics*. https://doi.org/10.1093/acrefore/9780199384655.013.423

Kelder J. M. and Waal W. J. I. 2019. *FROM 'LUGAL.GAL' TO 'WANAX': Kingship and Political Organisation in the Late Bronze Age Aegean*. Leiden: Sidestone Press.

Lazaridis I., Mittnik A. et al. 2017. Genetic origins of the Minoans and Mycenaeans. *Nature* 548:214–18. https://doi.org/10.1038/nature23310

Martirosyan H. 2013. The place of Armenian in the Indo-European language family: the relationship with Greek and Indo-Iranian. *Journal of Language Relationship* 10(1):85–138. https://doi.org/10.31826/jlr-2013-100107

Petrosyan A. Y. 2018. The problem of Armenian origins: myth, history, hypotheses. *Journal of Indo-European Studies* Monograph Series number 66.

Schrijver P. 2019. The case for Hatto-Minoan and its relation to Sumerian. In: *Talking Neolithic: Proceedings of the Workshop on Indo-European Origins Held at the Max Planck Institute for Evolutionary Anthropology, Leipzig, December 2–3, 2013*. Kroonen G., Mallory J. P. and Comrie B. (eds). 336–74.

Skourtanioti E. et al. 2023. Ancient DNA reveals admixture history and endogamy in the prehistoric Aegean. *Nature Ecology & Evolution* 7:290–303. https://doi.org/10.1038/s41559-022-01952-3

Smith A. T. 2015. *The Political Machine: Assembling Sovereignty in the Bronze Age Caucasus*. Princeton, New Jersey and Oxford: Princeton University Press.

Waal W. 2023. Deconstructing the Phoenician myth: 'Cadmus and the palm-leaf tablets' revisited. *The Journal of Hellenic Studies* 1–36. https://doi.org/10.1017/S0075426922000131

BIBLIOGRAPHY

Conclusion: Shibboleth

Atkinson Q. D. et al. 2008. Languages evolve in punctuational bursts. *Science* 319:588. https://doi.org/10.1126/science.1149683

Bilaniuk L. 2024. Memes as antibodies: creativity and resilience in the face of Russia's war. In: *Dispossession: Anthropological Perspectives on Russia's War Against Ukraine*. Wanner C. (ed). New York: Routledge. https://doi.org/10.4324/9781003382607

Blasi D. E., Michaelis S. M. and Haspelmath M. 2017. Grammars are robustly transmitted even during the emergence of creole languages. *Nature Human Behaviour* 1:723–9. https://doi.org/10.1038/s41562-017-0192-4

Bromham L. et al. 2022. Global predictors of language endangerment and the future of linguistic diversity. *Nature Ecology & Evolution* 6:163–73. https://doi.org/10.1038/s41559-021-01604-y

De Haas H. 2023. *How Migration Really Works: A Factful Guide to the Most Divisive Issue in Politics*. New York: Viking.

Elms T. after Tyshchenko K. 2008/1997. (On the lexical difference between Ukrainian and Russian.) See for example: https://alternativetransport.wordpress.com/2015/05/04/how-much-does-language-change-when-it-travels/ and https://blogs.cornell.edu/info2040/2018/09/17/38095/

Evans M. 2013. *The Language of Queen Elizabeth I: A Sociolinguistic Perspective on Royal Style and Identity*. Oxford: Wiley-Blackwell.

Hodges R. and Prys C. Number of Welsh speakers has declined – pandemic disruption to education may be a cause. *The Conversation*. 15 December 2022. https://theconversation.com/number-of-welsh-speakers-has-declined-pandemic-disruption-to-education-may-be-a-cause-196184

Léger G. 2016/2020. Le miroir d'un moment. https://gabrielleger.com/index.php/project/nn/2016-le-miroir-dun-moment/

Mackenzie, D. The truth about migration: how it will reshape our world. *New Scientist*. 6 April 2016. https://www.newscientist.com/

article/mg23030680-700-the-truth-about-migration-how-it-will-reshape-our-world/

Makarenko M. 1933. *The Mariupil burial-place: summary*. Kyiv: All Ukrainian Academy of Sciences. https://elib.nlu.org.ua/view.html?id=1044

McWhorter J. How immigration changes language. *The Atlantic*. 14 December 2015. https://www.theatlantic.com/international/archive/2015/12/language-immigrants-multiethnolect/420285/

Mufwene S. S. 2002. Colonisation, globalisation, and the future of languages in the twenty-first century. *MOST Journal on Multicultural Societies* 4(2).

Putin V. 2021. On the historical unity of Russians and Ukrainians. http://en.kremlin.ru/events/president/news/66181

Shnirelman V. A. 2012. Archaeology and the national idea in Eurasia. In: *The Archaeology of Power and Politics in Eurasia: Regimes and Revolutions*. Hartley C. W., Yazicioğlu G. B. and Smith A. T. (eds). 15–36. Cambridge: Cambridge University Press. https://doi.org/10.1017/CBO9781139061186.003

Trudgill, P. 2011. *Sociolinguistic Typology: Social Determinants of Linguistic Structure and Complexity*. Oxford: Oxford University Press.

Vince G. 2022. *Nomad Century: How to Survive the Climate Upheaval*. London: Allen Lane.

Zuckermann G. 2020. *Revivalistics: From the Genesis of Israeli to Language Reclamation in Australia and Beyond*. New York: Oxford University Press.

ENDNOTES

Introduction: Ariomania

1 There were earlier sorties from Africa of *Homo sapiens*, but the one that got underway sixty thousand years ago is thought to have been the one that founded all present human populations beyond that continent.
2 I'll use 'before the Common Era' (BCE) and 'Common Era' (CE) to refer to dates before or after year one respectively. These labels are used in exactly the same way as 'before Christ' (BC) and 'anno Domini' (AD) were in the past, in some parts of the world, but they are the secular equivalents. Where possible I will use the formula 'x hundred/thousand years ago', which I think is intuitively the easiest to understand since it uses the present as the point of reference, but it won't always be appropriate or practical. The first millennium BCE, which is sometimes called (no less accurately) the last millennium BCE, denotes the period 1000 BCE to 1 BCE. The first millennium CE is 1 CE to 1000 CE. By historical convention there is no year zero.
3 The *langue* (or strictly speaking *langues*) *d'oïl* is also alive, in the sense that at least one of them, modern French, is still spoken. *Oïl* became *oui*.
4 From Dante's *De vulgari eloquentia* ('On eloquence in the vernacular'),

published in the early 1300s, and translated from the original Latin by Nicholas Ostler. See Ostler 2005.
5 From William Cowper's *Retirement* (1782).
6 From 'Ode on a Grecian Urn' by John Keats (1819).
7 Historical linguists were once called philologists (they still are sometimes). There are small differences between the domains of historical linguistics and philology, depending on where you are in the world and how they are defined, but they are essentially the same.
8 From Mitchell's translation of *Beowulf*, published in 2018 by Yale University Press (New Haven, Connecticut and London).
9 A Balkan homeland was proposed by Russian linguist Igor D'Iakonov in the 1980s. Later that decade two other respected linguists, the Georgian Tamaz Gamkrelidze and the Russian Vyacheslav Ivanov, proposed a homeland in the Caucasus. See D'Iakonov 1984 and Gamkrelidze and Ivanov 1990.
10 The quotation comes from *The Rig Veda: An Anthology*, translated from the original Sanskrit by Wendy Doniger O'Flaherty (1981). Harmondsworth and New York: Penguin Books. All subsequent excerpts from the *Rig Veda* are taken from the same translation.
11 From 'Gypsy Song Taken From Papusza's Head' by Papusza (Bronisława Wajs), 1950/1951. English translation courtesy of the website of the Adam Mickiewicz Institute (culture.pl/en). The word 'gypsy', considered racist and pejorative by most Roma today, is derived from 'Egypt' – the wrongly presumed origin of the Roma in centuries past.
12 Like ships, languages are conventionally treated as female. Linguists talk about 'mother' and 'daughter' languages.
13 From 'Bog Queen' by Seamus Heaney, in the collection *North* (1975). London: Faber & Faber and in the collection *Opened Ground* (1998). New York: Farrar, Straus and Giroux.
14 See Sherratt 1988.
15 It may seem illogical to call the later incarnation 'Proto-Indo-European' when we know that there was an earlier ancestor, but there are two reasons for doing so. First, scholars disagree about how to label these two languages,

whose speakers' names for them are irretrievably lost, since they also disagree about how they were related (this disagreement is the subject of Chapter Three). Second, the later incarnation is the one on which most efforts of reconstruction have been concentrated. I will refer to the earlier language – the subject of Chapter One – as the 'lingua obscura'.

1. Genesis

1 There is one Eurasian steppe, but different sub-types within it that can be classified by climate or vegetation. Depending on the context, therefore, I will use the word in either the singular or the plural.
2 The rapids lay between the modern cities of Dnipro (the Ukrainian name for the river that flows through it) and Zaporizhzhia downstream (Zaporizhzhia, which means 'beyond the rapids', is the site of a major nuclear power plant today). They ceased to exist once the river had been backed up and raised to generate hydroelectric power in the twentieth century, through the construction of a series of dams including the Kakhovka Dam downstream of Zaporizhzhia. The destruction of that dam in June 2023, in the context of Russia's war on Ukraine, remodelled the landscape again, and unless or until it is rebuilt a version of the rapids could recreate itself. Though I'll use Ukrainian names for Ukrainian places in the main, I will stick to the better-known 'Dnieper' for this particular river.
3 The quotation is from a description by the Russian poet Mikhail Lermontov of a night-time journey through the Caucasus Mountains. It appears in his novel *A Hero of Our time* (1840) English translation by Nicolas Pasternak Slater 2013. Oxford: Oxford University Press. Reproduced with permission of the Licensor through PLSclear.
4 It was the Kuma-Manych Depression that had served as a conduit between the Black and Caspian Seas when their levels were stabilising. The chronology of its drying out is not well established, so just how wet the wetlands were at this date – and hence how much of a barrier they presented – is debated.

5 The brothers' names differ in the Qur'an, but little else does. The authors of those stories were themselves herders, explaining why their sympathies lay with Abel.
6 The culture is named for the first archaeological site at which it was identified, Baia-Hamangia in Romania.
7 There is an enchantingly large corpus of research on bone ice skates, including feet-on experiments. Ice skating was much more common prior to the Industrial Revolution, for everyday travel as well as for sport and fun. See Küchelmann and Zidarov 2005.
8 'Kurgan' is a Turkic word borrowed into English via Russian.
9 One offshoot of the original Lingua Franca survived until quite recently. Polari came to England in the sixteenth century with troupes of Italian performers, and was adopted as a secret code by the gay community when homosexuality was illegal there. 'Sea queens', gay men in the Merchant Navy, and drag queens spoke it. You might still catch a scrap of it in London's Soho, on nights when elderly gay theatrical types get together. 'Bona lallies ont the omi' is Polari for 'he's got nice legs'. See Baker 2002.
10 'Tell' is a word of Arabic origin that refers to an artificial mound or hill, often formed through the accumulation of debris. Words in other languages for the same concept, which crop up all the time in the archaeological literature, are *höyük*, *tepe* and *gora*. The first two are of Turkic origin, the third is Slavic.
11 This culture is commonly known as Cucuteni-Trypillia, or Trypillia for short (both before and after the birth of the megasites).
12 From *Jason and the Golden Fleece (The Argonautica)* by Apollonius of Rhodes, translated by Richard Hunter (2009). Oxford: Oxford University Press. Georgian researchers have linked the Greek myth to gold-panning in the Svaneti region of Georgia (ancient Colchis). See Okrostsvaridze, Gagnidze and Akimidze 2016.
13 The name of the oligarchs' culture is Maykop, after the site of the 'royal' kurgan that contained a man and two women.
14 Did this launch a long tradition of burying people with their favourite

vehicles? Scythian warriors were buried with their horses (a common practice among many steppe nomads). Wealthy Celts were buried with their chariots. Near the southern French village of Martignargues, no earlier than 1936, somebody went to the trouble of burying a Citroën 2CV ('deux chevaux'). It was driverless, but who could forget Sandra West, heiress to a Texan oil fortune, who in 1977 was buried in her nightdress in a powder-blue Ferrari?

2. Sacred Spring

1 From the poem *My Testament* (1845), translated from the original Ukrainian by John Weir.
2 The name Remontnoye is related to the English word 'remount', which in the past referred to the provision of fresh horses to soldiers to replace those that had been injured or killed in battle. Russian troops used to come to this area to buy remounts from the Kalmyks.
3 There were nutrients the Yamnaya's diet probably lacked, however, including folic acid, which is considered essential for the healthy development of the fetal brain and spinal cord. They would have known disability, then. We know hardly anything about their attitudes towards infirmity, other than that, if they spoke Proto-Indo-European, they had some words for it. The reconstructed Proto-Indo-European lexicon includes words for skin diseases, infertility, ophthalmological conditions (cross-eyedness, one-eyedness), deafness and lameness.
4 From 'The Stone Idol' by Ivan Bunin, translated from the original Russian by Irina Zheleznova. In: *Ivan Bunin: Stories and Poems* (1979). Moscow: Progress.
5 The oldest known saddle dates to about 700 BCE and was found in modern-day China. Stirrups that support the whole foot don't appear in the archaeological record until about a thousand years later, though whole-foot stirrups were preceded, in India, by a rope loop that supported the big toe.

6 The nobleman in question was Federico Grisone, and the quote is taken from his *Ordini di cavalcare* (*The Rules of Riding*), published to great acclaim in 1550. See *The Horse Magazine* 2018.

7 Sometimes I will give a stem, sometimes I will give the nominative singular form of it. A hyphen at the end will distinguish the former, but I won't observe the distinction strictly since sometimes the difference is minimal.

8 I'm using 'Proto-Indo-European' here as shorthand for 'speaker of Proto-Indo-European'. I'll occasionally use a similar shorthand for the speakers of other languages, and for the bearers of archaeological cultures (for example, 'the Yamnaya [people]' instead of the more unwieldy 'people associated with the Yamnaya culture'), but it is only a shorthand. As discussed in the Introduction, language, culture and genes contribute to identity, but identity is bigger and more complicated than any one of them.

9 Some Indo-European languages do without a word for 'one' entirely. If there is only one of something, after all, you hardly need to count it. Old Irish did have a word for 'one' (*oen*), but if a person wanted to say 'one cow' they would just say the word for 'cow': *bó*. *Bó ar fichit*, literally 'cow plus twenty', meant twenty-one cows. See Clackson 2007.

10 Latin *hostis* lives on in Spanish *hueste* and Romanian *oaste*, both of which mean 'army'.

11 These days, mythologists are often linguists by training. They use linguistic methods, notably the comparative method, to study myths.

12 An excerpt from the Roman historian Justin's summary of the *Philippic Histories* written by Trogus Pompeius in the first century BCE and later lost. Trogus Pompeius was the grandson of a Gaul. The Gaulish example is one given by Kim McCone, a Celticist at Maynooth University in Ireland who has lectured on the Indo-European concept of a sacred spring.

13 Adapted from Mallory 1989 with updates from Fortson 2010 and others.

3. First Among Equals

1 The theory that the scribe kept writing when he shouldn't have is American linguist Craig Melchert's. See Melchert 1991.
2 Cuneiform was a wedge-shaped script that was impressed on clay tablets using a stylus made from a reed straw. It was essentially a syllabary that also made use of ideograms. The people of Sumer in southern Mesopotamia invented cuneiform in the fourth millennium BCE, so Sumerian was the first language to be written in this script. Sumerian is a language isolate, meaning that, like Basque in Europe, it doesn't belong to any known language family. It was replaced by Akkadian, a language of the Afro-Asiatic family (Semitic branch), and both were deciphered in the nineteenth century. The same written code was therefore used to record very different languages, which is why it took so long for philologists to realise that Hittite belonged to a different family again.
3 Translated from the original German by Sophie Wilkins and Burton Pike (1995). New York: Alfred A. Knopf.
4 The tablet describes a religious ritual in Hittite, but at one point the writer states, 'And the priest speaks in the language of Kalashma in the following way . . .'. Fifteen to twenty lines of text follow in the previously unknown language. The city of Kalashma is mentioned in known Hittite texts, and was located in the north-west of Anatolia. The claim that the new language is Indo-European was made at a conference in September 2023, but at the time of writing that claim has yet to be published or subjected to peer review. Linguists Elisabeth Rieken and Ilya Yakubovich of the University of Marburg were due to publish a first linguistic analysis of it in *Archäologischer Anzeiger*, a journal of the German Archaeological Institute, in November 2024 (in German).
5 Old Europe's boundaries are a little vague. The term has been used at different times to refer to central and eastern Europe, south-eastern Europe and the Balkans.

6 Some researchers claim that the Anatolian ancestry carried by the Yamnaya was contributed by the farmers in the west, rather than having come through the Caucasus with the Iranian farmers. It's not impossible, in that time of broken borders – when steppe middlemen were plying their trade back and forth across the Pontic steppe – that both groups contributed. There might also have been contributions from the hunter-gatherers of Europe, who were displaced westwards towards the Dnieper by the arrival of the farmers.

7 The Leipzig database excludes almost all synonyms, for example. A given language will often have more than one word for the same thing (English is notorious for this), and different databases are more or less lax about how many synonyms they admit. This is a potential problem because a computer program will calculate different degrees of familial resemblance, and hence generate slightly different family trees, depending on which synonym is used. Heggarty and his colleagues therefore selected the most frequently used synonym in each case, letting only 'dog' in for that animal in English, for example, and excluding the less used 'hound'.

8 Some people call that parent Proto-Indo-Anatolian, to distinguish it from the Proto-Indo-European presumed to have been spoken by the Yamnaya. Others call it Early Proto-Indo-European, and the language spoken by the Yamnaya Late Proto-Indo-European.

9 The lack of farming-related words in the ancestor of Anatolian and Proto-Indo-European is also an argument against Colin Renfrew's Anatolian hypothesis, since that hypothesis holds that the speakers of the ancestral language lived in Anatolia and were among the inventors of farming.

10 These lines were translated by the American linguist Calvert Watkins from a Hittite text written in the seventeenth century BCE (the first line of the myth, quoted in Chapter Two, was also translated by him). If you hear an echo of the story of baby Moses in his reed basket, or Romulus and Remus being set adrift on the Tiber, or Sargan of Akkad on the Euphrates, you are not mistaken. In the

Near East at the time, and later further afield, there was much cross-fertilisation of mythologies (both within and between language families). See Watkins 2004.

4. Over The Range

1 The official name of Xinjiang today is the Xinjiang Uyghur Autonomous Region.
2 Brahmi was probably adapted by Indian scholars from a western alphabet brought to the subcontinent in the last centuries BCE by seafaring Semitic traders. Its descendants are still in use today. See Pillai 2023.
3 Tokharistan, which the ancient Greeks called Bactria, is now divided up between the nations of Uzbekistan, Tajikistan and Afghanistan.
4 Poem and translation abridged from Mallory 1989.
5 The description of the royal palace of Kucha comes from the seventh-century-CE *Book of Jin*, which was an official history of China published by the court of the Tang dynasty.
6 Some scholars have argued for the existence of a third language spoken along the southern rim of the desert, Tocharian C, but this is controversial.
7 The Proto-Uralic homeland is also debated. The family is named for the Urals, but its origins are now thought to have lain further east, perhaps in Siberia. A western branch gave rise to Finnish, Estonian, Saami and Hungarian.
8 Sogdian was the vernacular of Samarkand in what is now Uzbekistan, but Sogdian merchants were also the middlemen on the Silk Roads. They constituted a significant minority in the caravan cities at the edge of the desert.
9 Afanasievo is the name of a local mountain.
10 Ringbauer identified the cousins using a relatively new approach called 'identity by descent'. This detects long-range relationships (that is, relationships beyond the nuclear family) by looking at segments of

DNA that two individuals share. Because of the way that genetic material splits and recombines through procreation, the number and length of the segments is informative about the number of generations that separate the individuals from a common ancestor, up to about a dozen degrees of relatedness. See Ringbauer 2024.

11 See Cope 2014.

5. Lark Rising

1 One popular etymology of 'rune' starts with a reconstructed Proto-Indo-European word *h_3reuH- (bellow, roar), tracing its semantic development through 'lament', 'whisper, conspire' to 'counsel' but also 'secret'.
2 From *The Poetic Edda* translated from the original Old Norse by Edward Pettit. Cambridge: Open Book (2023). I have left out the question marks that Pettit inserted to indicate his uncertainty about the translations of some words.
3 One 2021 study found that the horses which the Yamnaya would have hunted or herded on the Pontic-Caspian steppe left a negligible trace in Europe's equine gene pool, suggesting that they didn't bring them west. That finding is disputed, however, with others reporting that European horses from this time on owed approximately a fifth of their ancestry to steppe horse lineages. Physical anthropologists, meanwhile, have detected horsemanship syndrome in a small proportion, three per cent, of European Yamnaya skeletons. See Librado 2021, Maier 2023 and Trautmann 2023.
4 At the time of writing, an unpublished report exists of a Yamnaya child buried further west than the River Tisza, in a cemetery otherwise filled with farmers. Such exceptions, while intriguing, are vanishingly rare.
5 The people associated with the Corded Ware and Yamnaya cultures shared about three-quarters of their DNA, but not their Y chromosomes. Some think of the two groups as related clans who separated

ENDNOTES

in the Pontic-Caspian steppe and followed different trajectories west. Others prefer the theory that Corded Ware people were descendants of the Yamnaya, and the debate has implications for how close the languages were that the two groups spoke. A study published in 2023 revealed that some individuals associated with Corded Ware were indeed direct descendants of Yamnaya people. Neither culture is fully visible to us, however. As discussed in Chapter Two, the burials that they left behind might, in both cases, have been those of an elite. See Ringbauer 2024.

6 From 'Flight', in *New and Collected Poems: 1931–2001* by Czesław Miłosz. Copyright © 1988, 1991, 1995, 2001 by Czesław Miłosz Royalties, Inc. Used by permission of HarperCollins*Publishers* and Penguin Books Limited.

7 The Amesbury Archer was buried around 2300 BCE. We know that he was a first-generation immigrant because analyses of the enamel on his teeth revealed that he had grown up in the Alps. The final additions to Stonehenge, including the construction of the famous avenue, are associated with Beaker burials, which suggests that they adopted the monument for their own religious purposes. See Armit and Reich 2021.

8 From William Shakespeare's *Cymbeline*.

9 Another possibility is that descendants of Corded Ware people who settled east of the Baltic – in the region that stretches from modern Lithuania up to Finland – island-hopped across that sea around 2000 BCE, bringing primitive Germanic dialects with them. In research that is unpublished at the time of writing, geneticists have reported a previously unsuspected but large wave of immigration via the sea to Jutland around that date (see McColl 2024). The linguistic evidence has to be reviewed in light of this discovery, but if Germanic arrived from that direction, and later than previously thought, it could explain early loans between the Germanic and Balto-Slavic tongues that have long puzzled linguists.

10 Gaul comprised modern France, Belgium and Luxembourg, along with parts of the southern Netherlands, south-west Germany,

Switzerland and northern Italy. In northern Italy, Gaulish replaced Lepontic in the last few centuries BCE.

11 Both Italic and Celtic form superlatives using derivatives of the Proto-Indo-European ending *-ismmo- (giving -ssimo in Italian, as in *massimo*, and *is* in Irish), while Germanic uses *-isto- (giving English '-est', as in 'biggest').

12 No known western Indo-European language converted the *e* of Proto-Indo-European **ped-* (foot) into an *a* (**ped-* became *podi* in Greek, *pēs* in Latin and *foot* in English), which is why the Dutch linguist Peter Schrijver has argued that these words came from one of the lost Indo-European languages of Europe.

13 Most linguists think that **tauro-* is Near Eastern in origin – a gift of the Neolithic farmers, that is. The Yamnaya may have had a different word for 'bull', which was replaced by **tauro-* when they came to Europe and encountered different breeds. Linguist Peter Schrijver has suggested an alternative possibility, however. According to him, **tauro-* may originally have come from a Caucasian language, and hitched a ride into Europe either with the farmers or with the Yamnaya themselves.

14 Some economists have suggested that the bust was an early example of 'Dutch disease', which is when the discovery of a new source of wealth – oil, say, or in this case metal – has a paradoxically negative effect on the broader economy in part because it leads to atrophy in other sectors. See Svizzero 2015.

15 The question of how violent the Yamnaya were is debated, however. While it's true that Yamnaya graves do not yield much evidence of violence as a rule, many of the Yamnaya skeletons excavated in the Lower Don region – close to the culture's presumed origins – do bear the scars of it. If the Yamnaya were the elite of a larger population, it is also possible that they delegated the fighting to subordinates. Most archaeologists feel more confident saying that people of the Corded Ware culture were violent, but not all. At a conference in Budapest in April 2024, Martin Furholt of Kiel University suggested,

provocatively, that the objects found in Corded Ware graves and conventionally identified as battleaxes might instead have been farming tools.

16 A study published in *Nature* in January 2024 showed that the Yamnaya carried genes into Europe that predispose their modern carriers to the inflammatory disease multiple sclerosis. In their prehistoric context, the authors suggested, those genes might have protected the Yamnaya against diseases of animal origin. See Barrie 2024.

17 Trautmann aired this theory at a conference in Poland in 2022. At the time of writing it has yet to be published or subjected to peer review.

18 The Y chromosome variant found most frequently among the Yamnaya was R1b-Z2103. The men associated with the Corded Ware culture carried mainly R1a, while a lineage of R1b unrelated to that of the Yamnaya, R1b-P312, dominated among Beaker males.

19 Historian Walter Scheidel has even suggested that the early Christians had nothing against polygamy. Augustine, an Algerian-born Berber who is remembered as one of the Fathers of the Church, distanced himself from monogamy by describing it as a 'Roman custom'. See Scheidel 2008.

20 This no longer holds in well-fed modern populations.

21 Genetic studies have forced a similar rethink with respect to historical peoples such as the Mongols. The impressive modern distribution of Genghis Khan's genes probably has less to do with his sexual appetite than with the combined reproductive efforts of his lineage over generations.

22 Ongoing excavations at the Tuscan spa town of San Casciano dei Bagni are revealing that wealthy Etruscans and Romans bathed side by side and may even have prayed together. There must have been tensions as Roman power grew, but the 'overthrow' of the Etruscans may have been less abrupt than was once thought.

23 A modern analogy is the way that, to the horror of purists, French has imbibed the English verbs 'to stop' (*stopper*) and 'to crash' (*se*

crasher) which it conjugates as it would any regular French verb. During the 2022 football World Cup, French commentators discussed how the objective of all the teams coming up against France was to *stopper Mbappé*, a reference to the star French forward Kylian Mbappé.

24 In fact this bilingual Latin- and Celtic-speaking elite began to emerge even earlier, after Caesar's brief forays into England in the middle of the first century BCE.

25 From Ramuz's *Riversong of the Rhone*, translated from the original French by Patti M. Marxsen. Blackhill, Australia: Onesuch (2015).

26 Geneticists have tracked Corded Ware ancestry from Jutland to Britain where, starting in the fifth century CE, it replaced much of the ancestry brought by the Bronze Age Beaker folk. See McColl 2024. The same, as yet unpublished study describes a medieval back-migration from Jutland towards Sweden and the Danish islands that could have implanted a Germanic language there. If true, this could explain why Vikings throughout Scandinavia referred to their language as 'the Danish tongue', since they perceived it as having come from present-day Denmark. The McColl et al. study also lends weight to the ancient oral traditions of the Goths and Lombards, according to which the former crossed the Baltic Sea from Gotland to Poland before heading further south, while the latter set off from the Jutland region to end up in Italy.

27 Dissection courtesy of Harvard Indo-Europeanist Jay Jasanoff, who added that it isn't always easy to distinguish between Latin and Latin-through-French. It was in a 1946 essay entitled 'Politics and the English Language' that Orwell lamented the corruption of English.

28 A comparison of Scottish and Welsh surnames reveals that Scots Gaelic is closer to Irish than it is to Welsh. Proto-Celtic *k^w became *p* in Welsh, but Irish retained the older derivative *k*, which it bequeathed to Scots Gaelic. Scots Gaelic *mac* ('son', as in Macduff or Macbeth) corresponds to Old Welsh *map*, which was often shortened to *ap* or even just to *p* or *b*. Thus Ap Rhys, 'son of Rhys', became Price, and Bevan means 'son of Evan'.

6. The Wandering Horse

1 From *Library of History* by Diodorus Siculus, translated from the original Greek by Charles Henry Oldfather. Loeb Classical Library. Cambridge, Massachusetts: Harvard University Press (1935). (And previous quotation.)
2 The name Harappan comes from the first such city to be excavated, Harappa.
3 *Assussanni* is derived from the Sanskrit *aśvasani*, meaning 'gaining or procuring horses'.
4 Early Indo-Europeans venerated cows, a crucial source of nourishment and wealth for them, but they had no objection to sacrificing or eating them. The forbidding of these things is thought to have postdated the Vedas. Modern Hindus also worship a different set of gods from their ancestors: Agni (Fire), Savitri (Sun), Varuna and Rudra receded over the millennia, and Shiva, Krishna, Ganesha and Kali, among others, came to prominence. Some gods remained constants, notably Vishnu, and the language of the hymns did not change very much.
5 From *Rubáiyát of Omar Khayyám*, translated from the original Persian by Edward FitzGerald. London: Quaritch (1859).
6 The Scythian and Sarmatian languages both belonged to the eastern branch of Iranic, like Pashto (and very unlike the western branch languages Persian and Kurdish, spoken south of the Caucasus), yet here they were naming rivers in the far west of the Iranic language space. River and place names aren't always reliable indicators of where ancient languages were spoken, in other words, especially when the speakers of those languages were highly mobile. See Heggarty 2015.
7 See Grünthal 2022.
8 Some Sintashta burials were ambiguous where gender roles were concerned, at least to modern eyes. Archaeologist Elena Kupriyanova of Chelyabinsk State University has reported cases where women were buried with weapons and chariot parts, and men were buried with female-associated ornaments. She thinks that rather than representing

a physical reality, these grave goods spoke to symbolic roles. For example, there might have been a social group associated with charioteering, and all those considered to belong to that group, including the relatives of the actual charioteers, might have been buried with its symbols. Those individuals buried with symbols usually associated with the opposite sex could, alternatively, have been shamans, since studies of traditional societies indicate that it wasn't uncommon for priests to cross-dress. See Kupriyanova 2023.

9 Salgado said this in Wim Wenders' 2014 documentary about him, *The Salt of the Earth*. The photograph in question is *The Gold Mine, Brazil*.
10 The link between sex or procreation and the extraction of metal persisted in many small-scale societies into the historical period. In some variations, women of reproductive age were excluded from the proceedings and smelting was the preserve of men who acted as 'midwives' to the metal. See Dupré and Pinçon 1997 (in French).
11 Strictly speaking, Andronovo is an umbrella term referring to a set of related cultures. The relationship between Andronovo and Srubnaya, and particularly between their languages, is still quite poorly understood.
12 The language spoken by the people of Gonur and neighbouring cities has been the source of much debate, however. The reason is that it is critical to determining which of the theories of the Indo-Iranian languages is correct. If you think that those languages crossed the Iranian Plateau more than five thousand years ago, then the people of Gonur might have spoken one of them. If instead you think that Indo-Iranian crossed the steppe later, then their languages were non-Indo-European and they first heard Indo-Iranian when the Andronovo rode through their gates.
13 The donkey was domesticated in Africa about seven thousand years ago, from where this new 'technology' diffused through Eurasia as far as China. The people of Gonur inherited the Semitic word for the animal from the Mesopotamians, and probably passed it on first to the Tocharians, and later to the Indo-Iranians. You can watch the

word evolve as it moves through the various speech communities: *ḫāru*, the Semitic word, became **koro* (mule) in Tocharian B and **khara-* in Proto-Indo-Iranian. See Kroonen 2018 (linguistic supplement to Damgaard et al).

14 From *The Histories* by Herodotus, edited by John M. Marincola and Betty Radice (consultant editor), translated from the original Greek by Aubrey de Sélincourt. London: Penguin (2003). All subsequent excerpts from *The Histories* are taken from the same translation.

15 The few Harappan burials that have yielded DNA amenable to analysis have mostly come from far-flung trading posts, including Gonur itself, and might not be representative, but they suggest that the Harappans carried a mix of indigenous hunter-gatherer and very ancient ancestry from the western end of the Iranian Plateau. It is because that Iranian ancestry predated agriculture, and the Harappans were farmers, that scholars infer that farming developed independently in India, through the experimentation of local hunter-gatherers. See Shinde 2019.

16 The Rosetta Stone, which was inscribed in Egyptian and Greek, allowed French philologist Jean-François Champollion to begin deciphering the Egyptian hieroglyphs in 1822.

17 The linguistic evidence suggests that the form of Sanskrit spoken in the Mitanni kingdom of Syria lacked the retroflex (called *murdhanya*, meaning 'in the head', by Sanskrit grammarians). It is present in East Iranic languages, but was probably borrowed into those from Indic.

18 Thracian is an extinct and poorly documented Indo-European language whose position within the family is unclear. It isn't the only one to fit that description. Others include Illyrian (once spoken in the western Balkans, up to the Adriatic coast) and Messapic, once spoken in the heel of Italy but apparently distinct from Italic and Greek.

19 In 1794, a bronze horn was retrieved from a bog at the foot of the hill on which the royal capital of Ulster once stood, in what is now Northern Ireland. The Loughnashade horn, as it became known, was roughly the same age as the Gundestrup cauldron, and it would have

been blown to summon Ulstermen to feasts. It is nearly two metres long and its end plate is decorated with writhing tendrils and lotus buds. The motifs are typically Celtic, but the lotus, *Nelumbo nucifera*, is not and almost certainly never has been native to the island of Ireland. Celtic art was inspired by Greek and Etruscan art, which was in turn influenced by civilisations further east.

20 From Calasso's *Ka: Stories of the Mind and Gods of India*, translated from the original Italian by Tim Parks. New York: Alfred A. Knopf (1998).

7. Northern Idyll

1 *The Balts* was published by Thames & Hudson. See Bibliography.
2 The Dnieper itself was named much later by Iranic-speaking Scythians, who were clearly present for long enough that their name for it caught on, but it may have had a Balto-Slavic name before that.
3 In 2018, geneticists reported a pulse of Corded Ware ancestry into the Baltic region much earlier, about 2800 BCE. This came as a surprise to linguists, since it predates the split of Baltic and Slavic and the arrival of the Balts in the Baltic, by their reckoning. They explain it, at present, by saying that the carriers of that ancestry almost certainly spoke an Indo-European language, but it wasn't Baltic. One proposal is that it was Germanic (which later island-hopped across the Baltic towards Jutland – see Chapter Five). See Mittnik 2018.
4 Strictly speaking, the parent of the national languages Finnish and Estonian, as well as of four smaller languages (Livonian, Vote, Carelian and Veps), is called (Proto-)Balto-Finnic, but I will call it (Proto-)Finnic here to avoid confusion with Baltic.
5 From the poem 'Steep Eyes of Wooden Gods' by Sigitas Geda, translated from the original Lithuanian by Jonas Zdanys. In: *Four Poets of Lithuania*. Vilnius: Vaga (1995).
6 A few runic inscriptions have been found in Novgorod.
7 See Carpenter 1981. It has been claimed that the dragon itself (as

opposed to its name) was inspired by Tolkien's traumatic encounter with flame-throwing tanks at the Battle of the Somme during the First World War.

8. They Came From Steep Wilusa

1 Linear A was so named because it consisted of abstract groupings of lines which, though derived from ideograms, were no longer recognisable as representations of objects.
2 Phrygian is the only one of the many dead languages in the so-called Balkan group that I'll discuss, because without it, it is impossible to explain the living ones (at least according to some theories).
3 By the turn of the Common Era the Greek dialects were in such intensive contact with each other – especially in the main ports of that island civilisation – that a new dialect emerged from the mix. This was Koine Greek, and it became the lingua franca of Greece, and indeed of much of the Mediterranean region, for centuries to come. Koine Greek is the main reason why the dialects of Greek did not become distinct languages.
4 From Homer's *The Iliad*, translated from the original Greek by E. V. Rieu. London: Penguin (1950).
5 *Meze* or *mezze* probably has multiple origins, but Turkic- or Greek-speakers, or both, might have borrowed an older Persian word meaning 'taste' or 'relish'.
6 The name of the egalitarian Caucasian culture is Kura-Araxes, after two local rivers (see Smith 2015). It was the expansion of Kura-Araxes groups, between five and six thousand years ago, that David Reich has suggested pushed early Anatolian-speakers west through the Turkish peninsula. See Chapter Three.
7 See Petrosyan 2018.
8 The ten per cent estimate is considered too low by some, however, because of the difficulty of distinguishing inherited words from loans in a language that was written down so late. By comparison, nearly

five hundred (fifty per cent) of the thousand most frequently written words in the *Oxford English Dictionary* are of early Germanic origin (Anglo-Saxon and Old Norse combined). With thanks to linguist Brian D. Joseph of Ohio State University and Philip Durkin, principal etymologist of the *Oxford English Dictionary*, for their input on this point. See Durkin 2014.

9 From 'What Are These Mountains Thinking About' (1962–4) by Ismael Kadare, translated from the original Albanian by Robert Elsie and first published in English in *An Elusive Eagle Soars: Anthology of Modern Albanian Poetry*. London: Forest Books (1993). http://www.albanianliterature.net/authors/modern/kadare-i/kadare-i_poetry.html

10 The phrase 'Potemkin village', from which 'Potemkin palace' takes its cue, refers to the fake portable villages that Prince Grigory Potemkin was rumoured to have assembled (and disassembled, and reassembled a bit further on) along the Dnieper to impress his lover, Catherine the Great, when she came to visit. The historian Simon Sebag Montefiore explains that the insult was unwarranted, since the villages were actually built. Ironically, 'Potemkin palace' might also turn out to be unfair. See Kelder and Waal 2019.

11 The Phoenicians actually took their name from that of the date palm. They were the date-eaters, or to be precise, the people who dribbled red juice while eating dates. It was Herodotus who connected the term *phoinikeia grammata* to the Phoenician alphabet. See Waal 2023.

Conclusion: Shibboleth

1 See Fabbro 2022.
2 In terms of vocabulary, Ukrainian and Russian differ by about thirty-eight per cent (Ukrainian is actually closer to Polish, on this measure, since they differ by thirty per cent). By way of comparison, French and Spanish differ by forty per cent and Spanish and Italian by thirty-

three per cent. Phonologically – that is, in terms of sound – the differences between Ukrainian and Russian are significant, which is why the shibboleth works. Regarding the word *palianytsia*, Bilaniuk explains, the *l* and *ts* sounds are softer in the Ukrainian pronunciation than in the Russian one. Unlike Russians, Ukrainians would also palatalise the *n*, producing an approximation of the *ny* sound in English 'canyon'. See Elms/Tyshchenko 2008/1997.

3 Perhaps a dozen major languages have earned the dubious honour of the 'killer' label, not all of them Indo-European.

4 From 'here yet be dragons' by Lucille Clifton. In: *The Book of Light*. Copyright © 1993 by Lucille Clifton. Reprinted with the permission of The Permissions Company, LLC on behalf of Copper Canyon Press, coppercanyonpress.org.

5 Migration statistics are always approximate, in part because entries are counted more assiduously than exits.

INDEX

Aegean Sea, 39, 43, 58, 146, 157, 223, 237, 242
Afanasievo culture, 131–3, 134, 135–40
Afghanistan, 8, 40, 132, 133–4, 189, 190, 191, 200, 271
Africa, 10, 50, 268; ancient exodus from, 9–10, 39, 83; clicks of bushmen, 11; emergence of *Homo sapiens* in, 9
Afro-Asiatic languages, 11, 23, 262
agriculture: ards (primitive ploughs), 79; climate event in late-third-century BCE, 152, 191–2, 197, 204, 216–17; farmer-herder cultural divide, 40–1, 53, 54–6, 59; farmers from Anatolia move west, 38–40, 43–6, 52, 99–100, 101–2, 136, 242; farming revolution expands geographically, 38–41, 43–6, 52, 98, 99–100, 101–2, 136, 242; farming revolution in Mesopotamia, 10, 38, 97; farming societies of the Balkans, 38–9, 48–9, 52, 53–6, 74, 98, 101; Harappans in India, 202; Iranian farmer ancestry, 38, 40, 56, 103, 104–5, 106, 110–11, 198; and Lech Valley study, 166–8, 169; Srubnaya culture, 197–8; in Tarim Basin, 134, 135; vocabulary of, 11, 38, 78, 79–80, 110–11, 156, 194–5, 199, 204, 225–6; wheatfields of western steppe, 64; and Yamnaya's expansion, 59, 79–80, 148–9, 150–1, 161, 162, 164–5
the Akkadians, 94, 116, 186, 191–2
the Alans, 191
Alaska, 11, 83
Albania, 243, 246, 251, 252; Albanian language, 8, 16, 31, 240, 241–2, 244, 251–2, 266
alcohol, 170–2
Aleut language, 11
Alexander of Macedon, 173, 185–6
alphabets, 12, 77†, 88, 145–8, 238; Cyrillic, 223–4; runic, 147, 153
Altai Mountains, 63–4, 128, 129, 130, 131–3, 134, 135, 139–40, 150, 193
Amazons (caste of female fighters), 86, 87
Amesbury Archer, 152
Anatolia (Turkish peninsula): Celtic settlement of, 173, 174; and genetic link with steppe, 105–6, 109; geography of, 37; Hattians in, 100, 112, 113–14, 156; infusion of Seljuk Turks, 266; Kanesh (modern Kültepe), 85, 87, 112–13, 114; Museum of Anatolian Civilizations, Ankara, 245–6; Mycenaeans in, 254–5; Neolithic farmers, 38–40, 43–6, 52, 99–100, 101–2, 103,

104–6, 112, 136, 242; Phrygians, 94–5, 240, 241, 245–7, 250; Reich's theory on migrations from east, 106–7, 109, 110, 111; Renfrew's Anatolian hypothesis, 99–101, 102, 108–9, 162 *see also* Hittite Empire

Anatolian languages, *90–1*; Hattian, 100, 112, 113–14, 156; Hittite, 16, 24, 25, 55, 77†, 93–6, 100, 110, 113–14, 117, 245; inflected character of, 96; laryngeals (vanished consonants), 24, 88–9, 96; Luwian, 96, 111, 115, 117; Lydian, 96, 111, 117; Neshili, 112; Palaic, 96, 115; Proto-Anatolian, 100–1, 106–7, 110–11, 112; Reich's work on, 105–7, 109, 110, 111; relation to Proto-Indo-European, 31, 88–9, 96, 99–101, 102, 105–7, 108–9, 110–11, 128; seen as oldest branch of Indo-European, 15–16, 31, 96; Sturtevant's theory on, 106, 109; westerly clustering of, 101, 111

Andronovo culture, 197–9, 201, 204

Anglesey, island of, 208–9

animals: association of dogs with death, 197; domesticated, 38, 66–7, 73–4, 77, 79; human diseases of animal origin, 163–4; taboo words for, 78; on Ukok Plateau, 139; wild, 47–8, 68, 139–40, 156, 197, 207

Anittas (Hittite king), 113

Anthony, David, 1–2, 48, 54–5, 73–4, 101, 105, 137, 139–40, 164, 197, 199

Apollo (Greek god), 164

Arabic, 22, 50, 89

Ararat, Mount, 107, 250

archaeology, 97–9, 245–6; alcohol as hard to detect, 170; Bell Beaker culture, 151–2, 157, 159, 160–1, 162, 163, 166, 171; 'dragonstones' in Armenian highlands, 248; focus on culture, 20–1, 30; Griffin Warrior, 243; Hittite royal archives discovered, 95; horse-riding as hard to detect, 73; Kargaly mines in Urals, 196–7, 244; at Lake Varna, 41–3, 45–7, 49, 52, 53, 54,

101, 106, 109, 150, 209; Lech Valley study, 166–8, 169; megasites on Neolithic steppe, 55–6, 58–9; 'new archaeology' after war, 98, 99; nineteenth/early twentieth century, 19, 25–6, 238, 239; sherds, 66; state-of-the-art science used in, 1–2, 26, 67, 68, 69–70, 102, 107–9; triangulation with genetics and linguistics, 20–1, 25–8, 30–1, 134 *see also* Corded Ware culture; Yamnaya culture

archaeozoologists, 37

Armenia, 108–9, 110, 247, 248, 249–51, 271; language, 16, 31, 76, 240–1, 246, 247–51

Arthurian legend, 178

Artificial intelligence, 267

Aryans, 19, 185, 191, 201–5, 207, 249, 265

Ashoka, King, 187

Assyria, 94–5, 112–13

Atlantis, lost city of, 18

Attila the Hun, 30

Australians, Indigenous, 19

Austria, 154, 157–8, 159, 229

Austronesian languages, 11

Avars, 148, 221

Avestan language, 16, 127, 189–90, 194

Azerbaijan, 271

Babylonia, 94–5, 115, 186, 250–1, 253

Bálint, Marianna, 148

Balkans region: 'Balkan' sub-group of languages, 240–55, 266; Celts move towards, 173; collapse of farming in Copper Age, 52, 53–6, 74, 98, 101; farming societies of, 38–9, 48–9, 52, 53–6, 74, 98; mining boom in, 45, 46–7, 106; Neolithic hunter-gatherers in, 37, 103, 105; and the Phrygians, 245, 246; Slav migrations into, 221, 222; Yamnaya migration to, 243, 246

Ballymena, County Antrim, 263

Baltic languages, 16–17, 77†, 213–20, 229, 266; Old Prussian, 218, 225,

INDEX

229; split from Slavic, 215, 220, 223, 226
the Balts, 171, 213, 214, 219–20, 225–32
'barbarian' term, 23
bards/poets, 8, 82, 84, 158, 208–9, 237–9, 240, 242, 244, 252–3, 254–5
Basque language, 18, 101, 146, 251, 252
Bede, Venerable, 176
bees, 78, 80, 110, 156
Bel (Babylonian tyrant), 250–1
Belarus, 214, 218, 223, 228
Bell Beaker culture, 151–2, 157, 159, 160–1, 162, 163, 166, 171
Bengali, 189
Beowulf, 8, 17
Bergquist, Anders, 207
the Bible, 14–15, 36, 41, 112, 176, 264
Bienvenue chez les Ch'tis (*Welcome to the Sticks*) (cult French film), 274
Bilaniuk, Laada, 264
bilingualism/multilingualism, 10, 51, 54, 172, 175, 225, 268, 274–5
biodiversity, 10
biologists, computational, 29
Black Death, 102
Black Sea region, 7, 8, 31, 41, 42–50, 51–7, 63, 94; deluge theory, 35, 36; and the Greeks, 36–7, 56, 111; initial D in names of rivers feeding, 37, 190–1; Reich's analysis of Neolithic populations, 103–6; today, 272, 273
boats, 7–8, 39, 47, 58, 111, 151, 152, 157, 218, 228
Bosporus, 35, 36–7, 271
Botai culture, 72, 79, 137
Boudica (Celtic queen), 158
Brahmi (Indian script), 123
Brazil, 195–6
Breton, 17, 178
Britain: arrival of Anglo-Saxons in fifth century, 176–7, 208, 223, 268; arrival of Celtic in, 159–61, 173; Beaker presence in, 152, 159, 163; linguistic impact of Vikings, 177; Norman French in, 177–8, 224; Roman arrival in, 146, 160, 174–5; steppe ancestry reaches, 162

bronze production, 57, 67, 80, 112, 139, 150, 153; trade network for, 153, 157, 158, 193, 195, 196, 198–9
Brown, Dorcas, 1, 197
Bryant, Edwin, 201–2
Buck, Carl Darling, 100
Buddhism, 123, 124, 125–6, 130, 187, 188, 272
Bulgaria, 35, 41–6, 52, 145, 246, 271; Bulgarian language, 17, 223, 251
Bunin, Ivan, 'The Stone Idol', 68
Burgas, city of, 271
Burgundians, 175

Calasso, Roberto, 208
cannabis, 67, 171
Cappadocia, 112–13
Carpathian Basin, 38–9, 173
Carpathian Mountains, 39, 40, 63, 71, 150, 162, 219
Caspian Sea, 35, 37, 40, 63, 65, 185
Catalan, 50
catfish, 37
Catherine the Great, 196, 207
Caucasus mountains, 10, 37, 40, 41, 56, 63, 104, 106, 111, 191, 247–8; Armenian theory, 241, 242; Caucasian oligarchs, 56–7, 58, 68, 105; Indo-European origin theory, 108–9, 110
Celtiberian, 154, 159, 174
Celtic languages, 8, 15, 17, 31, 127, 145–6, 154–61, 173–7, 266; in Britain and Ireland, 29, 146, 159–61, 173, 177, 178–9, 223, 268, 269; and emergence of English, 176, 177, 268
the Celts, 133, 164, 165, 171, 172–5, 176–7, 207
chariots, 74, 110, 115–16, 184, 188, 192, 194, 197, 198, 244
Charlemagne, 221
China, 16, 64, 83, 121–6, 129, 149, 188; Old Chinese, 130; and the Tocharians, 16, 124, 125–6, 130, 132, 134–5
Christianity, 14–15, 174, 223–4, 228, 229, 231–2, 250

Civilization (board game), 103
climate change, 10–11, 26, 41, 138, 272–3; from around 4200 BCE, 53–6; crisis in 2200 BCE period, 152, 191–2, 197, 204, 216–17; Santorini eruption (c.1600 BCE), 238
Colchis, 51, 56
Constantinople, 95, 223–4, 226, 228
Cope, Tim, 137–8
Copenhagen, University of, 131, 136, 163
copper production, 26, 42, 49, 52, 80, 153; and bronze, 57, 67, 112, 139; Bulgarian heartland of, 44–6, 48; Erzgebirge or 'Ore Mountains', 157; Kargaly mines in Urals, 196–7, 244
Corded Ware culture, 150–2, 153–4, 171, 198, 200, 216–17, 219, 242, 248–9; Bell Beaker encounter, 151–2, 157, 166; push back to the steppe, 192–4, 204; and the Yamnaya, 150–1, 162, 192, 242, 248–9
Cornish, 17, 176, 178
Covid-19 pandemic, 136
creoles, 267–8
Crimea, 175
Croatians, 246
Croesus, King of Lydia, 96
Cú Chulainn, 165
Cyril, Saint, 223–4
Czech language, 17, 222, 223
Czech Republic, 150, 157, 223

Dagestan, 65, 249
Dani, János, 148, 149
Danish, 175
Dante Alighieri, 13–14, 21, 175, 267
Danube, River, 37, 38–9, 43, 63, 149, 150, 157, 190, 243
Darius the Great, 185, 189, 199–200, 220
Dasyus people, 204
death/funerary practises: of Bell Beaker culture, 151, 152; Caucasian oligarchs' rites spread, 58, 68, 105; changes in steppe tribes' rites, 48–9, 50, 58; of Corded Ware culture, 151; cowrie shells, 134; cremation, 157–8; graves of female warriors, 86–7; Khvalynsk cemetery, 46–7, 104; the Mariupol cemetery, 259–61; Mycenaean, 243; shift from radical egalitarianism in Caucasus, 248; Sintashta, 184, 190, 194; 'Urnfield' cremation style, 158; use of red ochre, 67–8, 131, 150; Varna cemetery, 41–3, 46, 49, 52; Yamnaya, 59, 65, 67–8, 69, 70, 131, 132, 138, 148–9, 150, 194 *see also* kurgans
Diodorus of Sicily, 186
Dnieper, River, 37, 39, 55, 69, 79, 191, 193, 218, 219, 220, 262; Dnieper Rapids, 37, 48, 84, 103, 105, 216–17, 228
Dniester, River, 37, 54, 58, 191, 219, 220; Moravian Gate, 150
Dobruja Plain, 43
Don, River, 37, 51, 65, 79, 190
Dravidian language family, 205
Durankulak (site in Bulgaria), 45, 46, 47
Dusinberre, Elspeth, 245
Dutch language, 17, 78, 175, 176
Dzungarian Basin, 129, 130, 133, 135

Ebla, Syria, 111
Egypt, ancient, 12, 94–5, 115–16, 147
Elamite, 186
Elfdalian, 175
Elizabeth I, Queen of England, 274
English language, 11, 17, 19, 178, 251, 274; as analytic, 75; and the Celts, 176, 177, 268; and colonisation of Ireland, 178–9; *hw* sound, 78†; increasing varieties of, 268–9, 275; influence of Norman French, 177–8, 224; Middle English, 17; Multicultural London English, 275; Old English, 7, 17†, 17, 78†, 78, 81, 176–7, 178, 208; possible Tocharian loan word, 130; prohibited combinations of consonants, 222; schwa in, 89

INDEX

Enlightenment thinkers, 18–20
Ennius (Roman poet), 274
Epic of Gilgamesh, 36
Esfandiar (Persian hero), 84
Estonian, 18, 213, 225
ethnographers, 29
the Etruscans, 113, 145, 146, 147, 173, 265
etymology, 24
Eurasian Steppe: *chernozem* or 'black earth' belt, 64, 79, 148; climate crisis in 2200 BCE period, 192, 204; climate extremes on, 64, 138; dairying becomes common on, 79; eastern steppe, 64; extent/geography of, *60–1*, 63–4; Genghis Khan's journey, 137, 192; Hungarian *puszta*, 147–50, 161, 162; and Indo-Iranian/Balto-Slavic, 217, 218; Mongolian steppe, 72; plants, 49, 64, 67, 79, 171; Pontic-Caspian steppe, 13, 37, 51, 58, 71, 86, 87, 130, 164, 166, 241, 243, 248–9, 262; Sintashta horse sacrifices, 184, 190, 194; southern Russian steppe, 64–8, 85–6, 109, 110, 272; vast energy reserves of, 68; western steppe, 63, 64–8, 97, 217
European Union, 271
Evans, Sir Arthur, 239
Evans, Nicholas, 269

Faroese, 175
Farsi (Modern Persian), 16, 186
Faust legend, 45, 209
feminism, 101
Ferdowsi, *Shahnameh* (*The Book of Kings*), 190
Fergus mac Léti, King of Ulster, 84
Fertile Crescent, 38
Finnish language, 18, 213, 225, 225–8
First World War, 94–5
food/diet: and **ghostis*, 81–2; butter/olive oil divide, 71; early crops, 38, 111; and farming revolution, 10; and farming societies of the Balkans, 49; and herding, 40, 56; lactose intolerance/tolerance, 70–1, 152–3; plants with medicinal or psychoactive properties, 49, 64, 67, 171; and science of archaeology, 26; of steppe tribes, 40, 67, 68, 70; use of salt for conservation, 46, 54; vocabulary of, 11, 38, 78, 79–80, 110–11, 156, 194–5, 199, 204, 226; of Yamnaya, 67, 68, 70–1, 73, 151; Yamnaya consumption of milk, 67, 70–1, 73, 151
Frachetti, Michael, 138, 139, 140
France, 29, 39, 154–5, 159, 178
the Franks, 175, 221, 224, 268
French language, 25, 78, 89, 174, 177–8, 224, 268
Frisian, 175

Galatasaray (Turkish football team), 174
the Gauls, 173, 174; Gaulish language, 17, 81, 85, 87, 108, 154, 159, 174, 178
Gemini (constellation), 208
gender: and age-set concept, 86, 165; and Corded Ware migration, 151, 242; co-wife concept, 166–7, 169; in early Indo-European societies, 80–1, 166–8, 169; female warriors, 86–7, 147, 170; Gimbutas' ideas on Old Europe, 99, 101, 162; mitochondrial DNA, 28, 48–9, 55–6; and new societies in Caucasus, 248; polygyny, 168; studies of kinship and custom, 166–8, 169–70; women in Copper Age Black Sea region, 44, 49, 51, 55–6; and Yamnaya institutions, 165
genetics: and adult milk consumption, 70–1, 72; ancient DNA revolution, 1–2, 27–8, 49, 69–70, 102, 104, 242; and the Balkan languages, 242; Bell Beaker-Corded Ware encounter, 151–2, 157, 166; code of ancient diseases, 163; DNA from ancient horses, 72; genetic turnover in Europe (around 2000 BCE), 102, 105, 159, 162–3; genomes of Tarim

mummies, 133–4; Harappan DNA, 202; human genome, 27; landmark papers on Bronze Age migration (2015), 102, 105, 162; and migration, 27, 28, 38, 101–9, 131–7, 159–60, 162–4, 193–4, 198–9, 219–20, 242–3, 246–8; mitochondrial DNA, 28, 48–9, 55–6; Reich's work with DNA, 1, 104–7; three-way mix in modern Europe, 102, 152–3; triangulation with linguistics and archaeology, 20–1, 25–8, 30–1, 134; and writers from the classical period, 242, 252–3; Y chromosome, 28, 69, 87, 102, 152, 159, 162–3, 166, 219–20, 242, 247; Yamnaya and Afanasievo link, 131–3, 134, 135–40; Yamnaya DNA samples, 69–70

Genghis Khan, 137, 192

Gening, Vladimir, 183–4

Georgia, 46, 51, 56, 247, 271–2

the Germani, 146–7, 164, 171, 174, 207, 219; arrival in Britain in fifth century, 176–7, 208, 223, 268; confused with Celts in histories, 172; the *Männerbund*, 84

Germanic languages, 14, 15, 17, 18, 31, 145, 214, 222, 229, 266; as centum languages, 127; Dante's *jo* languages, 13; Low German, 176–7; northern, western and eastern branches, 175; Old High German, 93, 176, 221; *Poetic Edda*, 147, 175; Proto-Germanic, 153–4, 156, 175; relationship to Italic and Celtic, 155–7, 159; runic alphabet, 147, 153

Gimbutas, Marija, 97–9, 100–1, 102, 109, 150, 159, 162, 213–14, 215, 219, 231

Goa Portuguese Creole, 267

Gobi Desert, 122

gold, 42–3, 46, 56–7, 139, 195–6, 209

Gonur, ancient, 198, 199, 200, 204

Gordion (Phrygian capital), 245, 246

Gorny (steppe settlement), 197

the Goths, 220–1, 251; Gothic language, 17, 81, 175, 176, 220–1

Gotland, island of, 220

Gray, Russell, 107–9, 110

Greece, ancient, 13, 116, 145, 165, 168, 172, 173, 203, 253–5; age-set concept, 85, 87; and Black Sea, 36–7, 56, 111; Dark Ages of, 237–8; deities/myths, 7, 8, 56, 170–1, 208; Europe's first historians in, 154; post-Bronze Age population displacements, 247

Greek language, 15, 23–4, 31, 77, 78, 89, 266; alphabet, 145, 238; ancient literature, 8, 82, 84, 158, 237–9, 240, 242, 244, 252–3, 254–5; in 'Balkan' sub-group, 240, 242–4; and Indo-Iranian, 197, 240; inflected character of, 75; nouns ending in *s* in the singular, 214; oldest texts in, 25; origin on steppe, 242–4; Proto-Greek, 244, 249; as remaining a single language, 16, 243–4; spared by Romans, 173–4; written from right to left at first, 146

Greenland, 11

Grimm brothers, 21–2, 209

Gundestrup cauldron, 206–7, 209

Haas, Hein de, 273

Hadrian (Roman emperor), 25

Hafez, love poetry of, 190

Haitian Creole, 267

Hamangia culture, 43–5

Harappans, 187, 202, 204–6; language of as undeciphered, 203, 205

Hartwick College, New York, 1

Harvard University, 1, 104, 107, 109, 135–6, 247

Hastings, Battle of (1066), 172

Hattushili I, Hittite king, 93–4, 96, 113, 114–15, 239

Hattushili III, Hittite king, 116

Hawkes, Christopher, 98

the Hayasa, 249–51

Hayk (friendly giant), 250–1

Heaney, Seamus, 27

Hebrew, 14, 23, 266, 269, 270

Hecataeus of Miletus, 36

INDEX

Heggarty, Paul, 108
herding societies: of Altai Mountains, 131, 132–3, 134, 139–40; the Andronovo, 197–9, 201, 204; Aryan arrival in India, 200–1, 202–5; cultural divide with farmers, 40–1, 53, 54–6, 59; and first wheeled vehicles, 57–8, 78–9, 110; herding infiltrates the steppe, 40–1, 48; and human diseases of animal origin, 163–4; the Kalmyks, 65; levels of violence in, 29, 54–5; limited use of farming techniques, 56, 79–80; power shift from around 4200 BCE, 52, 53–6; Proto-Indo-Iranian as a language of, 194–5; shift from radical egalitarianism in Caucasus, 248; 'tebenevka effect', 73 *see also* Sintashta culture; Yamnaya culture
Herodotus, 13, 21, 129, 200, 246, 255
Heyd, Volker, 148, 162
Hill, Eugen, 222–3
Hindi, 16, 189, 205, 231
Hindu Kush, 16, 184, 187, 191, 200
Hinduism, 8, 188, 201, 202–3 *see also Rig Veda*; the Vedas
histories/chronicles, 25, 86, 172
Hittite Empire: Anittas' conquest of Hattush, 113; Bogazköy (now Boğazkale), 94, 95, 114–15, 245; demise of (1200 BCE), 116–17, 158, 237; earliest known texts, 25, 100, 113, 187; and Egypt, 115–16; first texts referring to, 100, 112; Hattian vocabulary in language of, 100, 113, 114; Hattusha (capital city), 94, 96, 100, 113, 114–15, 116–17, 173; and the Hayasa, 249–50; language, 16, 24, 25, 77†, 93–6, 100, 110, 113–14, 117, 245; language of deciphered, 93–4, 95–6, 245; origin myth of, 85, 87, 113–14; royal archives, 95, 100, 113, 116; story of Telipinu, 110; texts referring to Mycenaean civilisation, 254–5; war chariots, 74
Holy Roman Empire, 224

Homer, 8, 82, 84, 158, 237–9, 242, 244, 252–3, 254–5
Horsa and Hengist legend, 208
horses, 65, 71–4, 188, 208, 216; Rome's October Equus ritual, 207–8; sacrificing of, 73–4, 183–4, 190, 194, 207–8, 228; and Sintashta culture, 184, 190, 194, 197; and the Yamnaya, 67, 71, 72–3, 74, 79, 110, 131, 149–50, 194
Hortobágy National Park, Hungary, 147–50
Hrozný, Bedřich, 93–4, 95–6, 188, 245
Hubert, Henri, 174
Hungary, 38–9, 63, 137, 157–9; arrival of Magyars, 148, 224; Hungarian language, 18, 146, 148; Yamnaya migration to, 147–50, 161, 162, 243
the Huns, 30, 106, 191, 221
hunter-gatherers, 9, 39–41, 48, 102, 128; Eastern, 37, 38, 40, 103, 104, 105, 106; Western, 37, 38, 103, 104
Hurrian language, 112, 188, 199, 249

Iberia, 14, 151, 162, 174, 175
ice age, last, 3–4, 9–10, 31, 35
Iceland, 136, 147, 177; Icelandic language, 175
Idanthyrsus (Scythian king), 200
identity, concept of, 30, 263–4, 270; group membership in prehistory, 264–5; language as battleground, 275; nationalist 'Aryan' myths, 19, 98, 201; nationalist use of prehistory, 19, 98, 201, 265
Idrisov, Idris, 64, 65–6
imperialism, European, 8, 15, 121–3, 124, 133–4, 135, 201, 267–8
India, ancient, 7, 8, 19, 40, 110, 171, 264–5; arrival of steppe ancestry in, 200–1, 202–5; *ashvamedha* (horse sacrifice), 183–4, 207–8; Brahmi deciphered (1837), 186–7 *see also Rig Veda*; Sanskrit; the Vedas
Indic languages, 16, 18, 130, 184; and Andronovo expansion, 198–9, 204;

PROTO

dating of the Vedas, 25, 187, 189, 202; earliest recorded use of, 187–8, 191; Indian scholarship on, 201–2; layer of non-Indo-European loanwords, 203–4; route from homeland, 185, 191–5, 198–200, 201–4; as satem languages, 127; separation from Iranic, 199, 203–4

Indo-European languages, 7–8, 11, 18–20, *256–7*; and alcohol, 170; 'Balkan' sub-group, 240–55, 266; Bell Beaker-Corded Ware encounter, 151–2, 157, 166; centum–satem switch, 127, 155, 217, 218, 240; databases of shared/inherited vocabulary, 108–9; evidence of as intrusive to Anatolia, 100–1; first speakers of, 94–5; Gray's work on homeland of, 108–9, 110; Italo-Celtic and Germanic schism, 157; oldest inscriptions, 25; politicised debates over origins, 19, 98, 185, 201, 265; Renfrew's Anatolian hypothesis, 99–101, 101, 102, 108–9, 162; spread by principally social mechanisms, 166–72; as spread by small/temporary displacements, 38, 169–70, 273; spread of in Europe, 150–3, 156–70; 'steppe hypothesis', 68, 69–80, 87–8, 97–9, 100–1, 102, 109, 150, 162; twelve main branches of, 15–17, 23 *see also* Proto-Indo-European and entries for individual branches/languages

Indo-Iranian language branch, 16, 86, 127, 129, 184, 240, 266; four Aryan strongholds, 191; and Greek, 197, 240; Iranian Plateau theory, 185, 195, 201; northerly steppe route hypothesis, 185, 194–201, 203–5; Proto-Indo-Iranian, 194–5, 249, 265; route from homeland, 185, 191–200, 201–4; and the ruki rule, 218, 240; shared features with Balto-Slavic, 216, 217–18, 219; and Tocharian, 128–30, 135, 185, 187 *see also* Indic languages; Iranic languages

Indus Valley Civilisation, 12, 187, 201, 202, 204–6

infectious disease, 2, 29, 163–4, 169, 192; Plague of Justinian, 163, 221

Iranian Plateau, 38, 40

Iranic languages, 16, 127, 184, 190–1, 194, 199–200; the *Avesta*, 189–90; contact with Tocharian, 128–30, 135; Old Iranian, 129; Old Persian, 189; Persian poetic tradition, 190; and the Scythians, 86; separation from Indic, 199, 203–4

Iran/Persia: and 'Aryan' term, 19, 191, 201; Iranian farmer ancestry, 38, 40, 56, 103, 104–5, 106, 110–11, 198; and William Jones' ideas, 15, 107; languages of, 8, 15, 16, 129–30, 185, 186; Middle Iranian languages, 129–30; Old Iranian, 129; Old Persian, 189; Persian Empire, 8, 22, 185–6, 189, 190, 199–200, 249; and Romani, 22, 25, 129 *see also* Iranic languages

Ireland, 17, 29, 39, 84, 162; arrival of Celtic in, 159–61, 173; Beaker presence in, 152, 159, 160–1; the *fían*, 84; history since twelfth century, 178–9; *Kings' Cycle* of myths, 171; slaver colonies in Scotland, 178

Irish language, 84, 85, 179, 269; Ogam alphabet, 146; Old Irish, 161, 165

Islamic conquests of seventh century CE, 22, 189, 266

Isle of Man, 178

Israel, state of, 269, 270

Italic language family, 17, 31, 145–6, 222, 224; adoption of alphabet, 145, 146; as centum languages, 127; Dante's Tuscan dialect, 13; parting of ways with Celtic, 157–8; Proto-Italic, 154, 157–9, 168–9; Proto-Italic split, 158–9; relationship to Celtic and

334

Germanic, 155–7, 159; written from right to left at first, 146
Italy, 145, 154, 158, 173, 174, 177; Italian language, 13, 14, 50, 175
Ivanov, Ivan, 42, 43, 46

Jadwiga, Queen of Poland, 229
Jakob, Anthony, 214, 215
Jamaican Patois, 267
Jewish diaspora, 266
Jogaila, Grand Duke of Lithuania, 229
Jones, Janet, 71–2
Jones, Sir William 'Oriental', 15, 21, 107, 203
Justin (Roman historian), 173
Jutland peninsula, 147, 151, 153, 156, 176, 206–7

Kadare, Ismael, 'What Are These Mountains Thinking About', 252
Kadesh, battle at (1274 BCE), 115
Kakhovka Dam, 262
Kalmykia, Russian republic of, 65–6, 74, 272
the *Kamasutra*, 188, 190
Kanesh, city of (modern Kültepe), 85, 87, 112–13, 114
Karagash, 132, 137
Karakum Desert, 198
Karanovo (archaeological site), 52, 54
Kazakhstan, 64, 72, 132, 140
Kelder, Jorrit, 254
Kérberos (Greek hellhound), 197, 240
Khotan, city of, 125, 129–30
Khyber Pass, 203
Kiezdeutsch in Germany, 275
Kikkuli (Mitanni horse-trader), 188
Kizil, cave paintings at, 123, 133, 135
Kosovo, 243
Kōzui, Count Ōtani, 122
Kraków Cathedral, 229–30, 231
Kristiansen, Kristian, 151, 163
Kucha, city of, 125
Kuma-Manych Depression, 40
Kurdish language, 16
kurgans, 48, 49, 85–6, 131, 132, 138, 194, 248; of Caucasian oligarchs, 56–7, 58; of female warriors, 86–7; use of in Russo-Ukraine War, 272; of the Yamnaya, 59, 65, 67–8, 69, 70, 138, 148–9, 150

Ladino language, 266
Lang, Valter, 225
language, definitions/essential features: ability to discuss the hypothetical, 9; definitions, 12–13, 19; five main families today, 11; inflected-analytic contrast, 75, 96; modal verbs, 155; prohibited combinations of consonants, 222–3; stable core of the lexicon, 76–7; words with new additional meanings, 24, 78–9
language erosion/endangered languages, 269–70
Lapland, 225
Latin, 7, 14, 15, 17, 18, 29, 77, 81–2, 154, 158–9, 251; absorbing of Greek words, 174; and concept of 'slave', 168–9; decline/fragmentation of, 13, 25, 50, 268, 269; in early Roman Republic, 145–6; after fall of Western Roman Empire, 175; inflected character of, 75; nouns ending in *s* in the singular, 214; and runic alphabet, 147; spread during expansion of Rome, 173–4; and survival of Celtic languages, 174–5; survival west of the Elbe, 175; 'vulgar' form, 50, 175, 268
Latvian language, 16, 213, 216, 218, 229
Lazaridis, Iosif, 247
Lech Valley study, 166–8, 169
Léger, Gabriel, 265–6
Leibniz, Gottfried Wilhelm, 15, 21
Lemnos, island of, 146
Lepontic, 154, 159, 173
the Levant, 38, 103, 104, 105
Lingua Franca ('Language of the Franks'), 50
lingua obscura (language preceding Proto-Indo-European), 31, 51, 58, 106–7, 109, 110–11

linguistic diversity, 10, 11
linguistics, historical: accidental word resemblances, 23; and alcohol in prehistory, 170; and bard's stock phrases, 82; comparative method, 23–5; computers used for, 24, 107–9, 135–6, 160, 267; and 'controlled speculation', 29–30; Dante's evolutionary account, 13–15, 21, 175; emergence of dialects, 10, 19; English–Sanskrit word pairs, 23, 76; equation of flux of people with flux of language, 98; evolutionary approaches, 13–20, 21, 107; first human languages, 9, 10; Grimm's Law, 21–2, 29, 81; Indo-European as best-documented family, 20; William Jones' ideas, 15, 21, 107, 203; language as uncensored view of the past, 22–3; languages as never static, 273–4; laryngeals (vanished consonants), 24, 88–9, 96; Latin–Sanskrit word pairs, 15, 77; Leibniz's ideas, 15, 21; lingua francas, 50–1, 268; loans masquerading as inheritances, 108; loanwords, 22, 24–5, 81–2, 112–13, 114, 129–30, 155–6, 195, 203–4, 220–1, 225–6, 247, 251; the Neolithic as linguistic heyday, 10, 11; in nineteenth century, 21–2; 'open-syllable conspiracy', 221–3, 226; reconstructed lexicon as palimpsest, 80; reconstructing long-dead proto-languages, 23–4, 75–7; relative age of linguistic features, 23; and Renfrew's Anatolian hypothesis, 100, 101, 102, 108–9; ruki rule, 217–18, 240; and Russo-Ukrainian War (from 2022), 260–1; semantic leaps, 24; and slackening of lower jaw muscles, 153; speech sound laws, 21–2, 29, 75, 76, 107, 217–18; triangulation with genetics and archaeology, 20–1, 25–8, 30–1, 134; vertical bilingualism concept, 10; working backwards into prehistory, 100, 153

Lithuania, 97–8, 213, 218, 229, 232; Lithuanian language, 16, 76, 214–16, 221–2, 227, 229, 231, 232
Lombards, 175
Luther, Martin, 176

Magyars (steppe horsemen), 148, 224
Makarenko, Mykola, 259–60, 261
Malay, 268
Mallory, James, 84, 101, 105, 124, 134
mammoth steppe, 139–40
Mandarin Chinese, 11
Manx, 17, 178
Marathi, 189
Marinov, Raycho, 41–2
Mariupol, Ukraine, 259–61
Marmara, Sea of, 35
Max Planck Institute for Evolutionary Anthropology, Leipzig, 107
Mayflower, voyage of, 163
McLaughlin, Rowan, 160
Mediterranean Sea, 35, 39, 50
Meillet, Antoine, 214, 215
Melanesia, 10
Mesopotamia, ancient: climate crisis in 2200 BCE period, 191–2; farming revolution in, 10, 38, 97; first cities in, 56; flood myths, 36; formation of the first states in, 11; Uruk civilization, 56–7, 105, 116
metalworking, 151, 173, 193, 195, 197; Bronze Age mining boom, 153, 157, 196–7, 244; smelting, 44–5, 46–7, 50, 57, 67, 131, 150; Yamnaya, 67, 80, 131, 139, 150 *see also* copper production
Methodius, Saint, 223–4
migration: and ancient DNA revolution, 1–2, 27–8, 49, 69–70, 102, 104, 242; ancient move out of Africa, 9–10, 39, 83; Andronovo expansion, 198–9, 201, 204; Anglo-Saxon arrival in Britain, 176–7, 208, 223, 268; arrival of steppe ancestry in India, 200–1,

202–5; Bell Beaker-Corded Ware encounter, 151–2, 157, 166; into Bronze Age Greece, 242–4; and climate change today, 272–3; Copper Age movement between Europe and steppe, 46–51; debate over farmers from Fertile Crescent, 38, 101–2; eastern return in 2200 BCE period, 192–4, 204; epic treks in history, 136–40; after fall of Western Roman Empire, 102; farmers from Anatolia move west, 38–40, 43–6, 52, 99–100, 101–2, 136, 242; farming revolution's 'colonisation by leap-frog', 38; genetic traces in modern populations, 27, 101; as a historic constant, 275; as major motor of language change, 21, 28, 29, 101–9; and 'new archaeology' after war, 98, 99; as powerful force in prehistory, 28, 38, 101–9, 161–3; and science of archaeology, 26; Slav movement into Balkans, 221, 222–3; spread of Corded Ware culture, 151, 154, 157, 192, 193–4, 216–17, 242; and spread of Indo-European languages, 59, 68, 69–80, 97–109, 131–3, 151–2, 156–70, 176–9, 199–205, 218–23, 240–51, 262; Srubnaya expansion, 198; steppe corridor, 53, 55–6, 58–9; 'steppe hypothesis', 68, 69–80, 97–9, 100–1, 102, 109, 150, 162; Tacitus on, 111; today, 271, 272–3; varied linguistic impacts of, 266–7; western Eurasia's view of direction of, 192; of Yamnaya to the east, 71, 74, 102, 105, 131–3, 136–40; of Yamnaya to the west, 71, 74, 102, 105, 147–50, 161, 162, 164–6, 243, 246–7
military culture: age-set and warband concepts, 84–7; Andronovo, 198; and back-migration in 2200 BCE period, 192–4; Hittite, 113–16; Hun/Avar, 221; invention of chariot, 192; Mycenaean, 238, 244, 252–3, 254; Roman, 207–8; Sintashta, 194; Srubnaya, 197; Yamnaya, 165
Minoan Crete, 12, 238, 239, 242, 244
Minusinsk Basin, Siberia, 131
Mitanni kingdom, 188, 191, 199
Mitchell, Stephen, 17
Mongolia, 64, 137
Mongols, 66, 106, 137
Moravia, 223–4
Mormons, 136
Moscow, 64, 219, 225, 227
Mufwene, Salikoko, 268
Mughals, 203
Munda language family, 205
Murshili I, Hittite king, 115
Musil, Robert, *The Man Without Qualities*, 95
Mycenaean civilisation, 237–9, 242, 243, 244, 246–7, 252–5
Mykhailivka (site in Ukraine), 262
mysticism, 147
mythology: absent fertility god myths, 109–10; and alcohol, 170–1; ancient Greek and Roman, 7, 8, 56, 84–5, 86, 87, 109–10, 170–1; ancient Indian, 7, 8, 110; bands of brothers, 84–6; creation myths, 83–6, 87; dragons, 8, 83–4, 147, 230–1; Father Sky, 7, 81; female warriors in, 86–7, 147, 170; foster brothers in Europe, 165; Hittite, 85, 87, 110, 113–14; Irish *Kings' Cycle*, 171; mythologists' use of comparative method, 28–9, 83; Norse, 7, 86, 110, 147, 170, 175; notion of banished youth, 87; Prometheus, 171; Pygmalion, 124; the serpent in, 83, 230, 231; smith who makes pact with a devil, 45, 209; toolbox for reconstructing, 82–3; Trito (prototype dragon-slayer), 83–4; twin deities and horses, 208, 216

Nazi Germany, 19, 98, 201
Nebra sky disc, 157
Nero, Roman emperor, 158

Netherlands, 275
Niemen, River, 214, 218, 225
Niger-Congo languages, 11
Nikitin, Alexey, 53, 55, 191, 260–1
Niya, city of, 121–2
Norman French, 177–8, 224
Northern Ireland, 263
Norwegian, 175
Novgorod, 226–8
Nuristani languages, 16, 108, 184, 191, 200

Occitan, 14, 50, 252, 274
O'Connell, Daniel, 179
Old Assyrian, 112
Omar Khayyam, 190
onomatopoeia, 23
opium poppy, 171
Orwell, George, 178
Oscan, 17, 154, 158–9, 168–9, 173
the Ossetians, 191
Ostrogothic dialect, 175
Ottoman Empire, 95

paganism, 228, 229, 230, 231–2
Pakistan, 40, 191, 200
Pashto, 16, 191
Patterson, Nick, 136
peat bogs, 26
Persepolis, 185–6, 199–200
Persian Empire, 8, 22, 185–6, 189, 190, 199–200, 249
Peyrot, Michaël, 128, 134–5
Phoenicia, 254
Phrygians, 94–5, 240, 241, 245–7, 250
Pitman, Walter, 36
plague, 163, 164, 169, 192, 221
plurilingualism, 51
Plutarch, 86
Pochuyky (Ukrainian village), 53, 59
Poetic Edda, 147, 175
Poland, 219, 220, 229–30, 231, 262–3; Polish language, 17, 223, 224, 229, 262–3
pollen, 26
Pompeii (ancient city), 159
population statistics, 10, 11
Portugal, 17, 154

Possehl, Gregory, 205
pottery/ceramics, 44, 46, 55, 58
Proto-Afro-Asiatic, 78
Proto-Indo-European, 19; Anatolian languages' relation to, 15–16, 31, 88–9, 96, 99–101, 102, 105–7, 108–9, 110–11, 128; ancient Greek and Roman, 83–6; Black Sea cradle of, 7, 8, 31, 51, 52, 94; and creation myths, 87; and first wheeled vehicles, 78–9, 110; and gender roles, 80–1; as highly inflected language, 75, 96; *kwékwlos*, 78–9, 110, 127; moral justifications for raiding, 84; no metal words reconstructed, 80; phonetic symbols and diacritics, 76; pronunciation, 88–9; reconstructed words, 23–4, 75–7, 78–82, 110–11, 195; rich vocabulary of herding, 111; and social structures, 80–2; sounds, 23–4, 75, 76–7, 88–9, 153, 217–18; vocabulary of dairying, 78, 79; Yamnaya as possible radiator of, 59, 68, 69–70, 74–5, 79–80, 87–8, 106, 109, 110, 164
Proto-Saami, 225
Prussia, 229
Puduhepa, Hittite queen, 116
Putin, Vladimir, 265
Pylos (Mycenaean town), 243, 253

Qur'an, 41

radiocarbon dating, 26–7, 99, 137
Ramses II, Egyptian Pharaoh, 116
Ramuz, Charles Ferdinand, 175
Reich, David, 1, 103–7, 108, 110, 136, 247, 264–5
Renfrew, Colin (Lord Renfrew of Kaimsthorn), 97, 99, 101, 102, 108–9, 162
Renfrew, Jane (Lady Renfrew), 97, 98, 99
Rhine, River, 151, 152, 176
Rhône, River, 37, 175
Richard the Lionheart, 274

INDEX

Rig Veda, 82, 171, 183, 184, 189, 194, 203, 204–5, 239; the Ashvins (twin charioteers), 208, 216; date of, 25, 187, 189, 202; Indra (warrior-god), 19, 84, 87, 199, 204
Ringbauer, Harald, 135–6
road networks, 16, 57–8, 121–2, 125, 129–30, 157, 167–8, 196
the Roma, 22, 266
Romance languages, 14, 266; and Basque, 252; Dante's *langue d'oc* and *langue d'oïl*, 13–14, 175; Dante's *sì* language, 14, 175; hiving off from Latin, 25, 50, 268; and 'vulgar' form of Latin, 50, 175 *see also* Italic language family
Romani language, 16, 22, 25, 129
Romania, 17, 63, 175, 207, 224
Rome, ancient, 8, 29, 30, 145–6, 154, 168, 173–4, 207–8; age-set concept, 85, 87; arrival in Britain, 146, 160; colonies of North Africa, 50; deities/myths, 7, 17, 84–5, 86, 87, 170–1; Eastern Empire, 223–4; and the Etruscans, 113, 173; fall of the Western Empire, 102, 174, 175, 215; and fragmentation of Latin, 25, 50; the *luperci*, 84–5, 86
Rose, Brian, 245–6
Rostov Oblast, 66, 272
Rudra (Indic god), 164
Rumi, love poetry of, 190
Russia, 171, 196, 207, 220–1, 226–7, 228, 229, 271–2; nationalist use of prehistory, 265; southern Russian steppe, 64–8, 85–6, 109, 110, 272; Soviet Union, 65, 97–8, 213, 259–60, State Historical Museum, Moscow, 64–5
Russian language, 17, 223, 224
Russo-Ukrainian War (from 2022), 68–9, 260–1, 264, 271, 272
Ryan, William, 36

Sabines, 17, 87, 173
sacred spring (*ver sacrum*) concept, 87, 173
sacrifice rituals: of dogs and wolves, 86, 197; the Gundestrup cauldron, 206; horse sacrifices, 73–4, 183–4, 190, 194, 207–8, 228; human sacrifice, 56–7, 248; linking metal to human fertility, 197; and myth-making/storytelling, 83, 84, 189, 208; and Proto-Indo-European lexicon, 81, 83; Sintashta death rites, 184, 190, 194; steppe herds kept for, 40; and the Vedic poets, 189
the Saka, 190
Salgado, Sebastião, 196
salt production, 46, 54
Samoyedic languages, 128
Sanskrit, 7, 9, 15, 22, 78, 81, 93, 185, 187–90, 201–6, 244; association with the sacred, 7, 8, 16, 187, 188, 189, 202–3; and Avestan language, 189–90, 194; Brahmi script deciphered (1837), 187; inflected character of, 75; the *Mahabharata* and the *Ramayana*, 188, 190; Mitanni use of, 188, 199; nouns ending in *s* in the singular, 214; 'Out-of-India' theory, 18, 201, 203; similarities to Lithuanian, 216, 231; sounds, 23–4, 77†, 205–6; and Tocharian, 130
Santorini, 238
Sarmatians, 66, 191
Śárvara (Indic hellhound), 197, 240
satemisation, 127, 129, 155, 217, 218, 240
Saussure, Ferdinand de, 24, 96
Schliemann, Heinrich, 238
Schrijver, Peter, 160, 267
Scots Gaelic, 17, 178
Scott, James C., 29
Scott, Walter, *Ivanhoe* (1819), 177–8
the Scythians, 29, 53, 66, 86, 148, 190, 191, 192, 199–200, 203, 220
Second World War, 102, 262, 271
Seljuk Turks, 266
Semitic languages, 23, 112, 238, 254
Serbia, 243, 246
Serbo-Croatian, 223, 251
Serra Pelada gold mine, Brazil, 195–6

Shakespeare, William, 17
Shevchenko, Taras, 63
Shiraz, Persian city of, 208–9
Shishlina, Natalia, 64–6, 67, 68–9, 73, 79, 148
Shuppililiuma I, Hittite king, 115
Sicily, 145
Sieg, Emil, 123, 126
Siegfried (dragon-slayer), 84, 147
Siegling, Wilhelm, 123, 126
Silk Roads, 16, 121–2, 125, 129–30
silver, 206–7
Singh, Lalji, 264–5
Sinhalese language, 16
Sinop, 47
Sino-Tibetan languages, 11, 262
Sintashta culture, 184, 185, 190, 193–4, 195, 201, 217, 249; Arkaim site, 265; heirs of, 196–200, 201; and horses, 184, 190, 194, 197
Slavchev, Vladimir, 41–2, 43, 44, 52
slavery, 56, 153, 168–9, 178, 193, 224
Slavic languages, 23–4, 78, 214, 215, 216, 266; Cyrillic alphabet, 223–4; ejection of closed syllables, 221–3, 226; Gothic loans to, 220–1; nexus with Baltic, 16–17, 215, 218–20; Old Church Slavonic, 221–2, 223–4, 227; Old Novgorod, 227; ousting of Gothic by, 176; Proto-Slavic, 220, 222; shared features with Indo-Iranian, 216, 217–18, 219; split from Baltic, 215, 220, 223, 226; three-way divergence, 223, 224; and the Vikings, 228
the Slavs, 171, 176, 219–20, 221–3, 224, 226–7, 251
Slovakia, 223
Slovenian language, 17
Smith, Adam (archaeologist), 248
Smok (serpent denier-of-life), 230, 231
snakes, 78, 83, 206, 230, 231–2
social/political structures: borders imposed on dialects, 19; Caucasian oligarchs, 56–7, 58, 68, 105; charismatic leaders, 138–9; Copper Age societies of south-eastern Europe, 41–7, 48–50, 51–2, 53–6; growth of first states, 11; herder-farmer cultural divide, 40–1, 53, 54–6, 59; hierarchies, 56, 81, 116–17, 153; Hittite, 113–17; inequality and endangered languages, 270; nation-state embraced in eighteenth century Europe, 274; and Proto-Indo-European, 80–2; shift from radical egalitarianism in Caucasus, 248; of Uruk civilization, 56–7; Yamnaya institutions, 165–6
Sogdians, 16, 77, 129–30, 190
soma or *haoma* (intoxicant), 171, 199–200, 204
Spain, 145, 151, 154, 177
Spanish language, 252
Spondylus shells, 43, 45, 52
Srubnaya culture, 196–8, 244
Stavropol, 58
Stein, Marc Aurel, 121–3, 133
steppe tribes/societies: adoption of herding, 40–1; Andronovo expansion, 198–9, 201, 204; Caucasian oligarchs, 56–7, 58, 68, 105; chiefly elite, 48–51, 55; Copper Age trade network, 46–51, 54–5, 57–8; eastern return in 2200 BCE period, 192–4, 204; evidence of overgrazing, 192; forest-steppe, 53, 55, 57, 58–9, 128; and genetic link with Anatolia, 105–6, 109; horse-riding nomads of Bronze/Iron Age central Asia, 129; Iranic languages spoken by, 190–1; Khvalynsk, 46–7, 48, 52, 73–4, 84, 104; megasites of Copper Age, 55–6, 58–9; migration corridor, 53, 55–6, 58–9; and origins of Greek, 242–4; and Proto-Indo-Iranian, 194–5; southward flow of ancestry from 2500 BCE, 247–8; Srubnaya expansion, 198; steppe ancestry in India, 200–1, 202–5; 'steppe envoys', 47–51, 54–5, 74; and Uruk civilization, 56–7; and use of alcohol, 170; and Yamnaya's journey east, 137

INDEX

Stifter, David, 155, 174
Stockhammer, Philipp, 166, 169
Strabo, 86, 172
Strasbourg, Oaths of, 25
Sturtevant, Edgar, 106, 109
Sumer, 11, 12, 93, 191–2
Swahili, 268
Swedish language, 21, 175
Syria, 115–16, 187–8, 191, 199

taboo words, 77–8, 230–1
Tacitus, 111
Tajikistan, 190
Taklamakan Desert, 121–6, 134, 135
Tarim Basin, 121–6, 129, 130–1, 133–4, 135, 188
Taurus Mountains, 112
Taylor, Timothy, 207
Tbilisi, 271–2
Tell Brak, 56
Teutonic Knights, 229
Thangaraj, Kumarasamy, 264–5
Thessalonians, 223
Thracians, 207
Tibetan Plateau, 188
tin production, 112, 139, 153, 157
Tisza, River, 149, 150, 157, 162
Tocharian language, 15–16, 31, 123–6; as centum language, 127; entanglement with Uralic, 128–9, 130; and Indo-Iranian languages, 128–30, 135, 185, 187; journey to Tarim Basin, 126, 127–33, 134–5, 136–40; massive grammatical/phonological changes, 128, 130; oldest surviving documents in, 132; and Proto-Indo-European, 123–4, 126, 127–8, 130–3; Tocharian A and Tocharian B, 126, 127, 129, 130, 135
Tolkien, J. R. R., *The Lord of the Rings*, 8, 230
Trabzon, 46, 51, 58
trade and commerce: in ancient Greece, 145; Assyrian in Kanesh, 112–13; Baltic-Aegean routes, 157; beginning of silk trade, 135; in Copper Age, 43, 46–51, 54–5, 56, 57–8; crisis at end of Bronze Age, 116–17, 158; first wheeled vehicles, 57–8, 78–9, 110; and language, 13, 47–8, 50–1, 125–6, 129–30, 207, 226–7, 262; network for bronze, 153, 157, 158, 193, 195, 196, 198–9; and Sanskrit, 188; Silk Roads, 16, 121–2, 125, 129–30; and spread of disease, 169; Uruk network, 56–7; Viking, 228; Yamnaya, 67, 80
Trautmann, Martin, 164
Troy, 96, 254–5
Turkey, state of, 271
Turkic languages, 126, 135, 221, 266
Tushratta (Mitanni king), 188

Ukok Plateau, 139
Ukraine, 1, 39, 63, 218, 225, 228, 259–61, 262; forest-steppe boundary, 53, 55; megasites in, 55–6, 58–9; Proto-Slavic homeland, 220; Russo-Ukrainian War (from 2022), 68–9, 260–1, 264, 271, 272; Scythians in, 29, 53; wheatfields of western steppe, 64; Yamnaya archaeology in, 66, 68–70, 130–1, 260–1
Ukrainian language, 17, 78, 221, 223, 224, 264
Ulster, 207
Umbrian, 17, 154, 158–9
United Nations, 116
University of California, Los Angeles (UCLA), 98, 99
Ural Mountains, 71, 193–4, 195, 197–8; Kargaly mines in, 196–7, 244
Uralic language family, 18, 128–9, 130, 195, 213, 214, 225, 262
Urartians, 249
urbanism, early experiments in, 55–9, 116, 187; fortified oasis cities of Central Asia, 198–9, 200, 204
Urdu, 16, 189
Uruk civilization, 56–7, 105, 116
'utopia' concept, 140
Uyghur language, 122, 123, 126

Vandals, 175
Varna, city of, 35, 41–2, 271; ancient settlement at Lake Varna, 41–3, 45–7, 49, 52, 53, 54, 101, 106, 109, 150, 209
the Vedas, 16, 187, 188, 194, 199, 201, 202–3 *see also Rig Veda*
Venetic language, 154, 158
Ventris, Michael, 239, 253–4
Vercingeto*rix*, 156, 174, 178
Vikings/Norsemen, 7, 85, 86, 136, 177, 193, 226–7, 228; Old Norse, 175, 177, 228
Visigoths, 18; Visigothic dialect, 175
Vistula, River, 230
vocal tract, human, 9, 22
Volga, River, 46–7, 51, 65, 193, 219
Völkerwanderung, 221

Waal, Willemijn, 253–4
'Walloon' term, 177
Weinreich, Max, 19
Welsh language, 17, 30, 176, 269
wheeled vehicles, 57–8, 78–9, 110
woolly mammoths, 47–8, 139–40
writing: Balts and Finns take up, 227; *boustrophedon* phase, 146; cuneiform script, 94, 112, 186; emergence of, 12, 145–8, 153; European looting of Tocharian texts, 123, 124; first Celtic inscriptions, 146, 154; first preserved writings in Europe, 146; hieroglyphic systems, 147, 238; Hittite language deciphered, 93–4, 95–6, 245; ideograms, 12, 93–4; introduction to Anatolia, 112; Linear A, 238, 239; Linear B, 238–9, 253; oldest known system, 12; phonetic symbols and diacritics, 76; spread of in West in first millennium BCE, 145–8; syllabaries, 12; transliteration into Roman script, 88; undeciphered ancient scripts, 12, 203, 205

Yamnaya culture: adoption of fully nomadic lifestyle, 59, 66–8, 97; Afanasievo genetic link, 131–3, 134, 135–40; and alcohol/cannabis, 171; and Corded Ware culture, 150–1, 162, 192, 242, 248–9; and creation myths, 84; death/funerary practises, 67–8, 69, 70, 131, 132, 138, 148–9, 150, 194; food/diet, 67, 68, 70–1, 151, 192; and horses, 67, 71, 72–3, 74, 79, 110, 131, 149–50, 194; and the Mariupol cemetery, 261; migration east, 71, 74, 102, 105, 131–3, 136–40; migration west, 71, 74, 102, 105, 147–50, 161, 162, 164–6, 243, 246–7; and the Mycenaeans, 242, 243, 246; patriarch of first clan, 170; and Proto-Indo-European, 31, 59, 68, 69–70, 74–5, 79–80, 87–8, 106, 109, 110, 164; settlements on southern Russian steppe, 65–7, 69–70; size of men, 164–5; as skilled metallurgists, 67, 80, 131, 139, 150; social institutions of, 165–6; spread through steppe, 71, 97, 102, 262; and Tocharian, 130–3; vanishes from archaeological record, 152, 248–9
Yerevan, 247, 250, 271
Yiddish, 266

Zagros Mountains, 38, 40, 185
Zalpa, city of, 113, 114
Zarathustra, 8, 16; the *Gathas*, 189–90; Zoroastrianism, 189
Zdanovich, Gennady, 183–4
Zuckermann, Ghil'ad, 270